D K. Sinha is currently professor of mathematics and dean, faculty of science, Jadavpur University. He did his doctoral work on theoretical studies in electromechanical and magnetomechanical interactions relating to piezoelectric, electrostrictive, magnetostrictive and magnetoelastic materials. It earned him Ph.D. of Calcutta University in 1954.

Professor Sinha is a fellow of Institute of Mathematics and Its Applications (UK) and National Academy of Sciences of India. He was President of Mathematics Section of sixty-sixth session of Indian Science Congress in 1978. His present research activities are in the areas of piezosemiconductors (with plasma streams), thermopiezoelectricity and biopiezoelectricity, catastrophe theory, and bifurcation theory, on which he has written a number of papers and delivered lectures. He is coauthor of *A Treatise on Mathematical Theory of Electricity and Magnetism*.

CATASTROPHE theory and applications

Mathematics of Mathematical Sciences
a series of monographs on
topics of contemporary interest

Editor: D.K. SINHA
Department of Mathematics
Jadavpur University
Calcutta 700032, India

This series envisages to bring out monographs which bring to the fore the unifying character of mathematical sciences. It aims at projecting trends in the fields of mathematical psychology, mathematical biology, mathematical geology, computer mathematics, etc. The series will include in-depth studies in areas of mathematical concepts, techniques and modeling. It will cover qualitative studies in applicable aspects of mathematics to such areas as catastrophe theory, bifurcation theory, graph theory, fuzzy sets, nonnegative matrices, singular perturbation theory, pattern recognition, compartment analysis, combinatorics, automata theory, etc.

The series will prove useful to mathematicians interested in the applications of mathematics and to its researchers and users who seek to know about its uses.

Correspondence regarding contributions should be addressed to the Editor.

CATASTROPHE theory and applications

Edited by D.K. SINHA
Department of Mathematics
Jadavpur University
Calcutta 700032, India

A HALSTED PRESS BOOK

John Wiley & Sons
New York – Chichester – Brisbane – Toronto

To
Bibhutibhusan Sen
for whom even a
name was symbolic
of the spirit of
Mathematics

Published by Halsted Press
a Division of John Wiley & Sons, Inc., New York

Library of Congress Cataloging in Publication Data
Main entry under title:

Catastrophe, theory and applications.

Bibliography: p.
Includes indexes.
1. Catastrophes (Mathematics) I. Sinha, D.K.
(Dilip Kumar), 1940–
~~QA 614.58.C37 1982~~ 514'.7 81–13376
ISBN 0–470–27303–8 AACR2

Manufactured in India

FOREWORD

Mathematics has proved itself a valuable tool for carrying on sustained argument and for unifying points of view, both within its own domain, and in many applications in the physical sciences. How is it that it has no similar success in the social sciences? This, according to Professor C.W. Kilmister, is a fundamental puzzle. The start to solve the puzzle was made by von Neuman and Morgenstern some thirty years ago. Of the later developments the investigations of Pokropp on aggregation problem, of Atkin and his coworkers on urban structure and of Réné Thom, E.C. Zeeman and others on catastrophe stand out prominently. The last one, which forms the topic of this seminar, is in many ways the most versatile in the ways of its approach and applications, and most promising. The important theoretical aspect is the proof that all models of a general class involving three parameters can exhibit only one of seven elementary catastrophes. Much remains to be done in tackling the unanswered questions notably that of whether Thom's "General Class" of models really includes all those which we want to consider in the social sciences. As Professor Kilmister predicts "it is an approach that gives promise of development, which will certainly attach many in the next decade."

According to its author, catastrophe theory is "more than mathematics—it is a philosophy, a way of looking at and describing the world which then demands its own mathematics. It is 'morphology', shape described in a very general space of any number of variables." It encompasses a great deal, if not everything. Describing practical catastrophe theory the author touches on the problem of time, on irreversibility in thermodynamics, on the meaning of measurement, on problems of philosophy of science. The problem of arrow of time comes into catastrophe theory through the concept of stability.

Ilya Prigogine who is developing what is called *social thermodynamics* has been trying to connect nonequilibrium events and structure or order. From biology he knows that order and nonequilibrium phenomena are related. In fact, increased complexity leads to increased order, virtually a keynote of evolution. But where are the models in the physical world? The answer has to be in the entrophy aspects of physical states operating through time. The nonequilibrium processes lead to another form of becoming. To perform thousands of reactions which a biological molecule system does, a very precise organization is needed. Prigogine thinks that irreversible processes provide such a kind of self-organization.

That an organized structure could originate in physical chemistry is a new point of view. But the origin of organized structures comes beyond some critical distance from equilibrium—which has been mathematically demonstrated by Prigogine. The models in the physical world have been wanting until the discovery of alternating blue and red flashes in the Belousov-Zhapotinskii reaction, and of oscillations in the biochemical breakdown of sugar in cells.

Contemporary chemistry is studying dozens of reactions now known as bifurcation phenomena—the shift from one state or structure to another, out of what may be called chaos of molecular randomness. New types of organizations are being reported far from equilibrium. The scale involves millions and millions of molecules—'supermolecular', size up to centimetres and time sequence 100's of seconds—high organization out of molecular chaos. A holistic approach to chemical reactions, the system acts as a whole. In the existing successes the understanding of physical and chemical process we have too much idealized situations in order to arrive at simple workable solutions.

Classical physics started with the statement that given the laws of mechanics and the initial conditions one could predict what would happen in future or tell the past. But it is simply an unhappy picture. In a world of absolute prediction no new innovations are possible, no joy of new unexpectedness; probabilistic approach has come to the forefront. There are always fluctuations, instabilities, that would drive the system into new dimensions.

Many have been the attempts to reduce biology to physics and chemistry. But, it is the problem of organization which has been baffling. Now we realize that organization, or negentrophy can arise from nonequilibrium phenomena, a property of the biological system.

If I have gone astray and that too in a discontinuous manner, it is because my behaviour like that of any other of my kind is subject to the theory of catastrophe, which, of course, was the subject matter of the seminar. This book has essentially grown out of the deliberations in the seminar.

With these few remarks, I am happy to write foreword for the book, with the hope that its contents will not only enrich our fund of knowledge but will also improve our life and living.

28th May 1980
32 Jodhpur Park
Calcutta 700068

S.K. MUKHERJEE

PREFACE

René Thom's catastrophe theory, though controversial in initial years, has come to stay. It has not merely raked up our conventional approach to ideas of mathematics but has also brought about a reappraisal of notions and phenomena, particularly, in their qualitative aspects. The catastrophe machine (of E.C. Zeeman) has added significant dimensions to the theory in the arena of applications and applicability of catastrophe theory. Recent years have truly witnessed tremendous advances in applicable aspects of catastrophe theory in a wide variety of fields such as physical, social, life, engineering, earth sciences, etc. Hence the need of a seminar to deliberate mainly on these advances with a view to

(a) veering researchers in these fields to catastrophe theory and to the areas of its applications
(b) apprising researchers with recent trends and latest developments in applicable aspects of catastrophe theory.

The speakers and participants to this interdisciplinary seminar were drawn from researchers in catastrophe theory and in its diverse fields of applications. Professor E. C. Zeeman, FRS, of Warwick University, UK, whose work and exposé on catastrophe theory and its applications have become a legend, was the key speaker. This volume is an outgrowth of the proceedings of the seminar held at Jadavpur University in 1979 and is intended, as it ought to be, to give a permanent shape to some facets of activity in the seminar. For the final touches of lecture notes of Professor Zeeman (which were taken by D. Sinha), we had to draw upon the classic collection of his and his coworkers' papers. The debt of gratitude to Professor T. Poston and Professor I.N. Stewart is acknowledged for using their excellent bibliography on catastrophe theory. My former student Subir Roy volunteered to prepare the index for the book. I am grateful to him for his valuable help. This publication would not have been a reality had we not been able to get the funds from Department of Science & Technology, Government of India, and thanks are due to them. This is an opportunity to acknowledge our thanks to University Grants Commission, Jadavpur University and British Council without whose help and assistance the seminar would not have taken place. We would deem our labours amply rewarded if the objectives set above are achieved.

Jadavpur University
Calcutta 700032

D.K. SINHA

CONTENTS

CONTRIBUTORS

1. A.K. DAS, Mathematics Department, Jadavpur University, Calcutta
2. P.J. HARRISON, Statistics Department, University of Warwick, Coventry
3. J.N. KAPUR, Mathematics Department, Indian Institute of Technology, Kanpur
4. C.K. MAJUMDAR, Physics Department, Calcutta University, 92 Acharya Prafulla Chandra Road, Calcutta
5. S. MITRA, Geological Sciences Department, Jadavpur University, Calcutta
6. T. POSTON, Battelle Memorial Institute, Battelle
7. A.K. RAY, Mathematics Department, Jadavpur University, Calcutta
8. A.B. ROY, Mathematics Department, Jadavpur University, Calcutta
9. DILIP SEN, Physics Department, Burdwan University, Burdwan
10. D.K. SINHA, Mathematics Department, Jadavpur, University, Calcutta
11. J.Q. SMITH, Statistics Department, University College, London
12. I.N. STEWART, Mathematics Institute, University of Warwick, Coventry
13. E.C. ZEEMAN, Mathematics Institute, University of Warwick, Coventry

ONE

Introductory Glimpses of Catastrophe theory

E.C. ZEEMAN

Catastrophe theory is a new mathematical method for describing the evolution of forms in nature. It was created by Réné Thom. It is particularly applicable where gradually changing forces produce sudden effects. These effects are often called catastrophes, because the intuition about the underlying continuity of the forces makes the very discontinuity of the effects so unexpected, and this has given rise to the name. The theory depends upon some new and deep theorems in the geometry of many dimensions, which classify the way that discontinuities can occur in terms of a few archetypal forms; Thom calls these forms the elementary catastrophes.

A cusp catastrophe is one of such elementary catastrophes. The cusp catastrophe can be derived from the quartic

$$V(a, b, x) = \tfrac{1}{4} x^4 - ax - \tfrac{1}{2} bx^2 \qquad (1)$$

having one behaviour variable x and two control parameters a and b.

The behaviour surface is given by

$$\frac{\partial v}{\partial x} = 0 \Rightarrow x^3 - bx - a = 0 \qquad (2)$$

and the fold curve is given by

$$\frac{\partial^2 v}{\partial x^2} = 0 \Rightarrow 3x^2 - b = 0$$

Bifurcation set, i.e. the hypersurface obtained by the projection mapping onto the control space is obtainable by eliminating x from (1) and (2) giving

$$27\,a^2 = 4\,b^3$$

THE BASIC CLASSIFICATION THEOREM

Let ρ be a 2-dimensional control (or parameter) space, let χ be a 1-dimensional behaviour (or state) space, and let f be a smooth generic function on X parameterized by C. Let χ be the set of stationary values of f (given by $\partial f/\partial x = 0$, where x is a coordinate for X). Then M is a smooth surface in $C \times \chi$ and the only singularities of the projection of M onto C are fold-curves and cusp catastrophes.

THE SEVEN ELEMENTARY CATASTROPHES

Let us suppose that C was 3-dimensional and χ was 1-dimensional. Then $C \times \chi$ would be 4-dimensional and M would be a 3-dimensional manifold in $C \times \chi$. A new type of singular point would appear called the swallow tail catastrophe. We can make the dimensions of χ to be greater than or equal to 2 and keep C 3-dimensional and obtain different types of singularities. These features are as shown below.

Dim. of C	1	2	3	4	5
No. of Catastrophes	1	1	3	2	4
Names	Fold	Cusp	Swallow tail, hyp. umbilic	Butterfly par. umbilic	— —

Example: Buckling of structures is a field obtaining many elementary catastrophes. This can be inferred from the fact that the equation of elastic structures capable of large geometrical changes under conservative loading can be characterized by variational principles containing non-quadratic functionals of the displacement.

Among the possible control parameters are load, imperfection, position of unassigned load, initial structure geometry and moduli. If we can decide that only at most two of these are 'essential', then we can apply the cusp catastrophes. For example, for a straight strut under compression by an assigned axial load, it is known that load and imperfection are essential control parameters. Here, the potential energy emerges precisely in the form of cusp catastrophe.

The cusp catastrophe shows that the five qualitative features of (i) bimodality, (ii) inaccessibility, (iii) sudden jump, (iv) hysteresis, and (v) divergence. All these are interrelated and the deep classification theorem permits us to enunciate the general principle that whenever we observe one of these five qualities in nature, we should then look for the other four and if we find them we should check them whether or not the process can be modelled by the cusp-catastrophe.

Catastrophe Machine

This is a small toy made out of elastic bands designed to illustrate catastrophe theory, and in particular, the cusp (or Riemann-Hugoniot) catastrophe. It is a concrete example in which all the variables are obvious and measurable. It gives a powerful intuition of how a continuous change in control can cause discontinuous jumps in behaviour.

Materials needed: Two elastic bands, three drawing pins, a piece of card board and a piece of wood.

The length of the unstretched elastic bands is taken to be the unit length. With this unit the wood needs to be about 2×6. A cardboard disc of unit diameter is cut. A drawing pin is used to pin the centre of the disc to the centre of the wood, so that it spins freely. A washer may be used for this purpose. Y, another drawing pin, is placed into the wood at a distance Z from X. The elastic bands are then fixed to the disc at a point near the circumference with the help of the third pin Z (Fig. 1). Now the other end of the elastic bands is slipped over Y, and the machine is complete, ready to go.

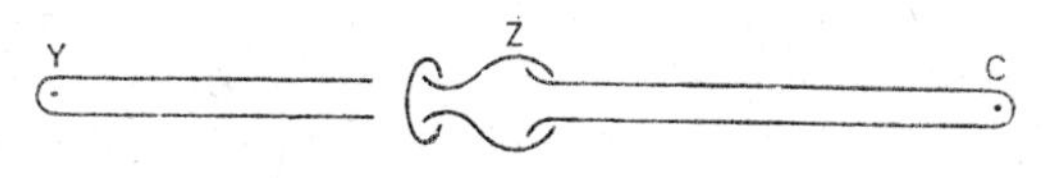

Figure 1

Holding the other end C of the other elastic band so that both elastics are taut, C is moved smoothly and slowly about the plane. The disc responds by moving smoothly and slowly most of the time, but occasionally it makes a sudden jump. This jump is called a catastrophe. The positions of C are marked each time the machine jumps. The positions will be found to build up a concave diamond-shaped curve with cusps P, P', Q, Q' (Fig. 2). This curve is called the bifurcation set B. The curve is symmetrical about the line XY. The cusps P, P' lie on the line XY at distances approximately $XP = 1.41$, $XP' = 2.46$.

It will be observed that the machine does not jump when C enters B, but only where C exits from B, and then only provided C has previously entered B from the opposite side. If C makes a complete circle round B, then the disc executes a smooth circle. If C makes little circles round P or P' then the disc includes once each time round. By contrast if C makes the circles round Q, Q' then the disc makes smooth oscillations. All these qualitative behaviours are surprising at first, but becomes easy to understand with the help of catastrophe theory. Therefore, let us put the mathematics into the framework of catastrophe theory.

Let,

Control point $= C =$ the held end of the elastic

Control space $= C =$ plane $=$ all possible positions of C

State, $\theta =$ angle $Y\hat{X}Z =$ position of disc

State space, $S =$ circle $=$ all possible positions of disc

Potential $V_c(\theta) =$ P.E. in elastic bands with control held at C, and disc held at θ

Therefore

V_c: $S \to R$ is a smooth function from the circle S into the reals R, and

V: $C \times S \to R$ is a smooth function $C \times S$ into the reals

If C is held fixed and the disc is released then the disc will jump into a state θ_c, i.e. a local minimum of V_c and stay there. Friction damps out any oscillations, so that the system is dissipative. Therefore the machine obeys the fundamental requirement of catastrophe theory, that the state θ seeks a local minimum θ_c of the potential V_c. It transpires that if C lies outside B then V_c has one minimum (and one maximum). Therefore the position of the disc is determined uniquely. However, if C lies inside B then V_c has two minima (separated by two maxima). The choice of which of the two minima that potentia V_c will take, and therefore which of the two positions the disc will seek, is determined by the past history of C, as follows. If C moves smoothly then a minm. of V_c will move smoothly, unless it happens to be annihilated by coalescing with a maximum as C exits from B, in the case when the disc were in the minm. that was annihilated, then it will have to jump into the other minm. This is illustrated in Fig. 3 which shows a sequence of potentials V_c as C runs along the dotted line in Fig. 2 near the cusp P, parallel to the b-axis. In each case the graph is drawn for small value of θ.

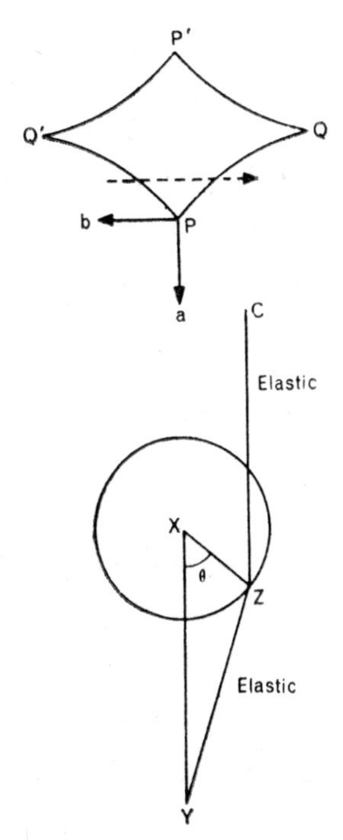

Figure 2.

The next diagram, Fig. 4, shows the graph of θ_c as a function of C drawn for small values of control coordinates $C = (a, b)$ with origin at the cusp P, and for small values of θ. More precisely the graph is given by $\partial V/\partial\theta = 0$, because the graph includes not only the minima

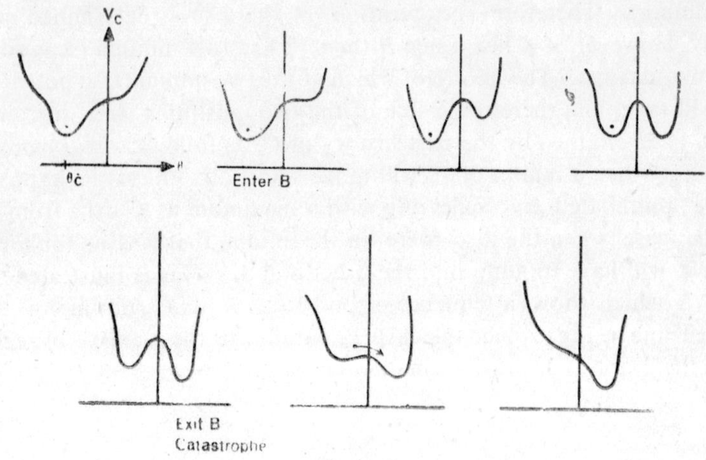

Figure 3

of V_c but also the maxima of V_c near $\theta = 0$, shown shaded. The other single valued surface of maxima near $\theta = \pi$ is not shown in Fig. 4, but is shown in Fig 5.

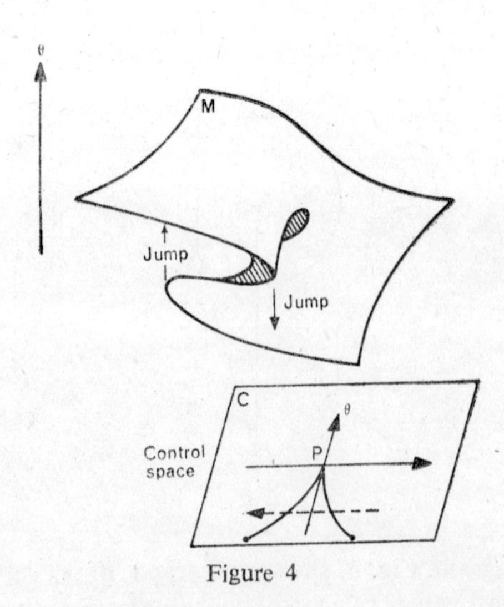

Figure 4

Using Hooke's Law for energy in the elastic bands, we can compute the approximate equation for $M: \theta^3 - 1{\cdot}3b\,\theta^2 + 1{\cdot}8\,a\theta + 1{\cdot}3b = 0$,

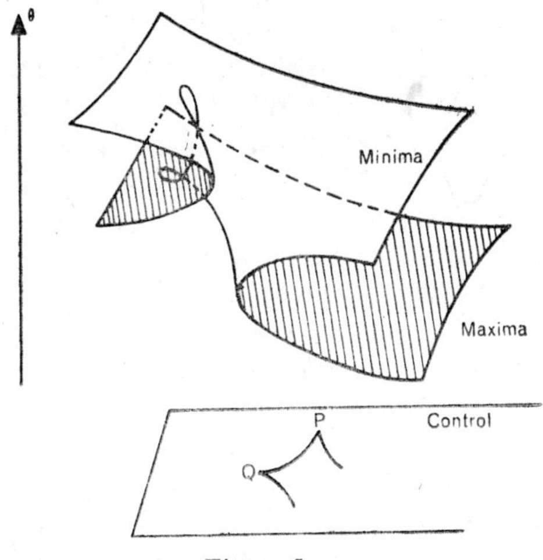

Figure 5

which is equivalent to the canonical cusp catastrophe surface $\theta^3 + a\theta + b = 0$. As C varies the state θ_c stays on the surface M of minima. Therefore as C moves along the dotted line the state has to jump from the lower branch of the surface on to the upper branch at the second crossing of the cusp. If C moves to and fro the state performs hysteresis action.

We may summarize Figs. 3 and 4 by saying that the machine obeys the delay convention which stays delay jumping until necessary. Delay is a consequence of the fact that the dynamics of the damped disc obeys a differential equation in θ with the property that θ and $\partial V/\partial\theta$ have opposite signs. An application of catastrophe theory satisfying a differential equation with this property will exhibit delay—for example, the application to heart beat and nerve impulse. By contrast the application to phase-transition does not exhibit delay. The reason is that the variables temperature, pressure and density in transition do not obey a differential equation direct but are averaging devices exhibiting the average behaviour of very many differential equations obeyed by particles of the substrates.

Again, in respect of the cusp θ a surface similar to Fig. 4 will be found, but dual in the sense that maxima and minima are interchanged. Putting the pictures for P and θ together, Fig. 5 will be obtained. It is difficult to combine pictures for all the 4 cusps, because this would involve bending the δ-axis round in a circle.

The precise shape of the bifurcation set B, checked by computer, has been studied by Poston and Woodcock. It will be as shown in Fig. 6. Here the lines represent the intersections of M with the planes θ = constant, and hence B is the envelope for all such lines.

Let us now turn to another area of application of catastrophe theory.

Konrad Lorenz in his book 'On Aggression' says that rage and fear are conflicting factors influencing aggression. The question is: Can we represent this similar sentence by a similar graph? Let us think the case of a dog. Another problem arises. Can we measure the rage and fear driven in a dog at any moment? Lorenz suggests that we can and he proposes that rage can be measured by how much the mouth is open and fear by how much the ears lay back.

Let us assume that the rage and the fear can be plotted as two horizontal axes, α and β. Also, let us assume that we can devise some vertical scale representing the resulting behaviour of the dog running from fight to flight, through intermediary behaviour such as growing, neutral and avoiding. We want to plot the graph x as a function of α and β. It is true that, as before, an increase in rage causes an increase in aggression and an increase in fear causes a decrease in aggression. But what if we increase both rage and fear together? The least likely behaviour is for the dog to remain neutral, and the most likely behaviour is fight or flight, although which of the two he will choose may be unpredictable.

Let us imagine a likelihood distribution for the behaviour x in each of the following four cases.

Drives	*Most likely behaviour*
1. Rage only	Fight
2. Fear only	Flight
3. Neither	Neutral
4. Both rage and fear	Fight or Flight

The interesting case is case 4 where the distribution is bimodal. What catastrophe theory tells us that if the likelihood distributions look as in Fig. 3, the graph will look like cusp, with catastrophe surface M pictured as in Fig. 4.

The curve on the surface where the upper and the lower sheets fold over into the middle sheet is called the *fold curve*, and the projection of this down into the horizontal plane C is called the bifurcation

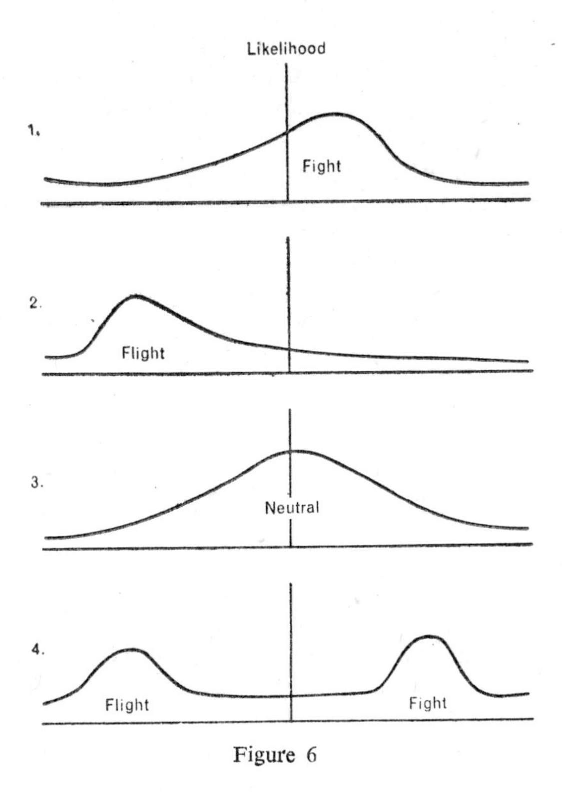

Figure 6

set. Although the fold curve is a smooth curve, the bifurcation set has a sharp point, forming a cusp, and this is the reason for the name *cusp-catastrophe*.

The surface gives us a new insight into the dog's aggression mechanism. For, as his drives vary over the horizontal plane C, so his mood and behaviour will follow suit over the surface M above. Path P_1 begins with the dog frightened, cowering in a corner, say. If we increase his rage, e.g. by approaching him too close in his territory, then his mouth will begin to open, but he will remain cowering until the point Q is reached. At that moment, he reaches the fold curve at the edge of the lower sheet, and so the stability of fleeing frame of mind breaks down, and he will suddenly catastrophically jump up into the upper sheet into a fi : ting frame of mind. Consequently, he may suddenly attack. Conversely, suppose he is in a fighting frame of mind and we cause him to follow path P_2 by increasing his fear in some way, then he will nevertheless remain in a fighting frame of mind until the point Q_2 is reached, when he will suddenly and catastrophically jump down onto the lower sheet into a fleeing frame of mind. Consequently, he may suddenly retreat.

The Machanism of Aggression Argument

Let us suppose that the model is describing the emotional mood of an opponent in an argument. The behaviour axis runs as follows:

Fight:	hysteria
	abuse
	irrational argument
	rational discussion
	concessions
	apologies
Flight:	tears

If we begin to make him angry and frightened, then we shall first deny him access to rational thought and force him to jump between irrational argument and concessions. If we make a little more so, then we shall next deny him access to those behaviour modes, and force him to make bigger jumps between abuse and apology. Finally, we make him very angry and frightened, then we shall limit his possible behaviour to only hysteria or tears. If our purpose is to persuade him to something that will make him both angry and frightened, the best policy is to state the case and go away, for our absence will allow his anger and fear to subside and give him access to ration thought again, and so enable him to understand our point.

TWO

Applicable Catastrophe Theory I stability of ships

E. C. ZEEMAN

Catastrophe theory provides a new way of looking at the statics of a ship, and this, in turn, lends a new simplicity of the global nonlinear dynamics. The weight of the ship and the position centre of gravity are taken as parameters. Then the set of equations form a smooth manifold that maps onto the parameter space. It is the singularities of the map that are recognizable as elementary catastrophes.

For example, heeling and capsizing are fold catastrophes. At the metacentre, there is a cusp catastrophe. The point of inflexion of the lever arm is caused by another cusp catastrophe. The increased likelihood of capsizing when overloaded, or when the crest of a wave is amid ships, is due to a swallow tail catastrophe. The evolution of hull shape from cause to modern ship is characterized by a butterfly catastrophe. On the metacentre locus, there are hyperbolic umbilic catastrophes. The suddden onset of heavy rolling due to nonlinear resonance with the wave is a dynamic field catastrophe.

Linear Theory or Rolling

We confine ourselves to the 2-dimensional problem of rolling only.

G C.G. of the ship

B_0 C.G. of water displaced

B_θ Centre of buoyancy when the ship is at angle θ

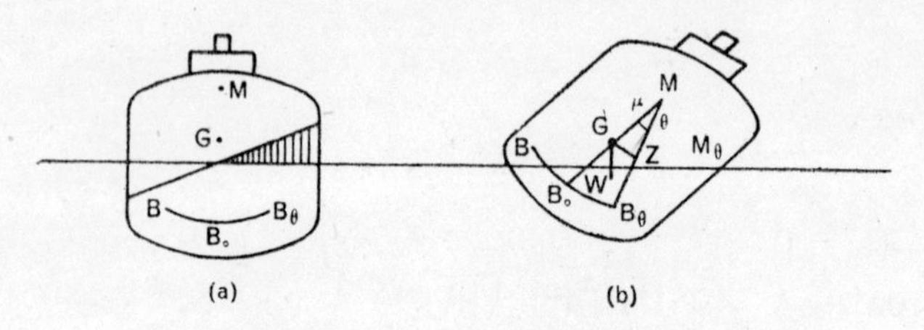

Figure 1. The buoyancy locus and metacentre.

B Buoyancy locus $= \{B_\theta, -\pi < \theta \leqslant \pi\}$

N Normal to B at B_θ

M Metacentre = Centre of metacentric height

Lemma. *B is a convex closed curve. When the ship is hulled at an angle θ, the normal N_θ is vertical.*

Corollary. *For small θ, the buoyancy force passes approximately through the metacentre M.*

The couple in Fig. 1b consists of weight W of the ship acting downward at G, and the buoyancy force W acting upward at B_θ. Let l denote the lever arm of this couple:

$$\begin{aligned} l &= \text{distance from } G \text{ to } N \\ &= GZ \end{aligned}$$

where Z is the foot of the perpendicular from G to N_θ. Newton's Law of motion gives

$$I\ddot{\theta} = -Wl \tag{1}$$

where I is the moment of inertia of the ship about G. From the corollary to Lemma 1 and Fig. 1b, we obtain

$$\begin{aligned} l &= \mu \sin\theta \text{ to second orders in } \theta \\ &= \mu\theta, \text{ again to second orders} \end{aligned}$$

Hence, the approximate linear equation is

$$I\ddot{\theta} = -W\mu\theta \tag{2}$$

or, $\theta = \theta_0 \cos \frac{2\pi t}{T}$, θ_0 = amplitude

T = period of the rolls

and $$T = 2\pi \sqrt{\frac{T}{W\mu}} \tag{3}$$

Considering the damping factor, the equation becomes

$$\ddot{\theta} + 2\epsilon\dot{\theta} + \frac{W\mu}{I}\theta = 0 \tag{4}$$

where ϵ is a constant.

This has the effect of multiplying solution (3) by a dacay factor $e^{-\epsilon t}$.

QUANTITATIVE ESTIMATES

In Fig. 2 (modified version of Fig. 1a),

let $2a$ = beam of ship = width at water line

A = area below water line

$x, y)$ = coordinates of B_θ relative to B_0

$P = B_0M$ = radius of curvature of B at B_0

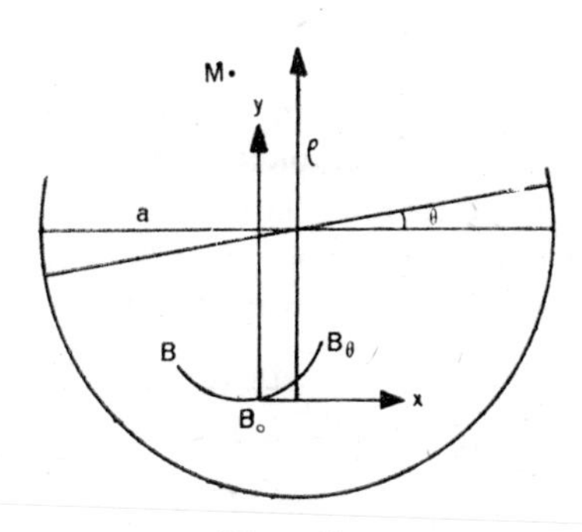

Figure 2.

Lemma. $\rho = 2a^3/3A$

Proof. Let $t = \tan\theta$. Let θ^n denote order t^n. Each wedge has area $a^2t/2 + O^2$, and the coordinates of the centre of gravity of the wedges relative to the mid-point of the waterline are $(\pm 2a/3 + O'$, $\pm at/3 + O^2)$. The coordinates (x, y) of B_θ are given by taking moments of the wedges about B_0.

$$A_x = \left(\frac{1}{2}\, a^2 t\right)\frac{4a}{3} + O^2 = \frac{2a^3 t}{3} + O^2$$

$$A_y = \left(\frac{1}{2}\, a^2 t\right)\frac{2\,at}{3} + O^3 = \frac{a^3 t^2}{3} + O^3$$

Putting $\rho = \frac{2}{3}\frac{a^3}{A}$ $x = \rho t + O^2,\quad y = \frac{1}{2}\rho t^3 + O^3$

Therefore equation of B is

$$y = \frac{x^2}{2\rho} + O\,(x^3)$$

which has the radius of curvature ρ at the origin, as required.

WALL-SIDED SHAPES

A ship is said to be wall-sided if the sides are parallel at the waterline. Let β denote the maximum angle of heel for which the waterline still meets the parallel part of the sides.

Lemma. *In a wall-sided ship, the buoyancy locus for* $|\theta| < \beta$ *is precisely the parabola* $x^2 = 2\rho y$

Now where the ships tend to differ in the height of G above the waterline, let us work out a complete example of a destroyer and a liner in order to illustrate the contrast. In such a case, we assume typical values for the beam as position of G, and deduce the resulting metacentric height and period of roll.

TABLE 1

		Destroyer	
Assume	*Beam*, $2a$	10 m	30 m
	Ht. of h above the waterline	0	2 m
Deduce	Metacentric ht	0, ρ m	0.5 m
	Period of roll	8 sec	30 sec

Greater metacentric height of the destroyer gives a greater couple, and makes her more stable, so that she can perform tighter maneuvres, as well as causing a faster roll.

Rolling Seaway

Typical Atlantic Ocean waves usually have a wavelength between 50 and 100 metres. Figure 3 shows a typical Atlantic wave of period 6 sec.

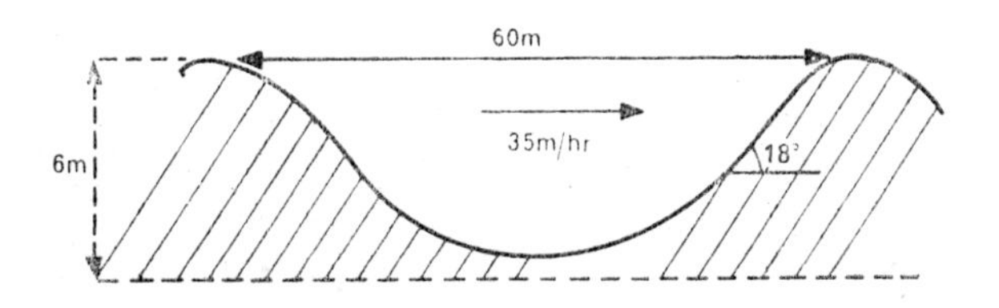

Figure 3.

Effects of the Waves on the Ship

Let, in our 2-dimensional example, the water surface be inclined at an angle α to the horizontal. The buoyancy force is the sum of all the pressures on the hull that would be required to keep the displaced water in equilibrium with the surface inclined at an angle α; hence the buoyancy force acts at $B_\theta + \alpha$ perpendicular to the water surface. Then the equation for rolling would be

$$I\ddot{\theta} = -W\mu\,(\theta + \alpha)$$

Let us now suppose that waves are coming from the side with period τ and maximum inclination α_0 to the horizontal. The inclination α will be a periodic function of time t, approximately equal to

$$\alpha = \alpha_0 \cos \frac{2\pi t}{\tau}$$

Thus, $$\left(\frac{T}{2\pi}\right)^2 \ddot{\theta} + \theta = -\alpha = -\alpha_0 \cos \frac{2\pi t}{\tau} \qquad (1)$$

Particular integral (P.I.) of (1) of period τ will give the rolling induced by the waves. The P.I. will be an amplification of the wave

$$\theta = \lambda\,\alpha = \lambda\,\alpha_0 \cos \frac{2\pi t}{\tau} \tag{2}$$

where λ is a constant amplifying factor. We can compute λ by substituting (2) in (1) such that

$$\left[-\left(\frac{T}{\tau}\right)+1\right]\lambda = -\alpha$$

Therefore $$\lambda = \left[\left(\frac{T}{\tau}\right)^2 - 1\right]^{-1} \tag{3}$$

We now compute the effect of a typical Atlantic beam sea upon our destroyer and linear. Assume $\tau = 6$ sec, and $\alpha = 18°$

TABLE 2

	Destroyer	*Linear*
Period of roll T, by Table 1	8 sec	30 sec
Amplifying factor	98/7	1/24
Amplitude of inclined roll	23°	1°

CUSP CATASTROPHE AT THE METACENTRE

Define the *metacentric locus* $\mathcal{M}$ of the ship to be the locus of centres of curvature of buoyancy locus; in other words, $\mathcal{M}$ is the evolute of B (Fig. 4).

Now B_0 is a point of symmetry of B where the radius of curvature is stationary, and hence M is a cusp point of $\mathcal{M}$.

Figure 4.

The geometric question arises—which way does the cusp branch, upwards or downwards? Figure 4 gives the answer and we have

Lemma. *The cusp branches upwards in the ship* (a) *and downwards in* (b).

We will now consider another question in regard to the physical significance of which way the cusp branches.

Question. Given the position of G, at what angle can the ship float in equilibrium?

Answer. It will be those values of angle θ, such that G lies on the normal N_θ to B at B_θ. But the normals to B are tangents to M. Therefore the angles are obtained by drawing tangents from G to the cusp.

[Ref: Fig. 11, p. 460, Zeeman, 1977.]

[In Fig. 11, we plot the graph θ as a function of G, for the two boats pictured in Fig. 10. In each case, the centre of gravity is represented by a point on the vertical axis R, and for simplicity, we assume $|\theta| < \beta$ where β is some suitable bond. On the vertical line above each position of G, we plot the corresponding values of θ, and as G varies these points trace out a smooth surface, which we call the equilibrium surface E.]

defn.:
$$E = \{(G, \theta) \colon G \in N_\theta, |\theta| < \beta\}$$
$$= \{N_\theta \times \theta; |\theta| < \beta\} \subset C \times R$$

Therefore E is a smoothed ruled surface, consisting of horizontal planes to the normals, one for each θ.

In other words, E is the normal bundle of B. If E is projected down onto C, it becomes folded, called the fold curve F, which projects onto the cusp. Hence the cusp is a bifurcation set. If, further, G lies inside the cusp when $\theta = 0$, equilibrium is represented by a point on the middle sheet of E, whilst the other two equilibria are represented by the points on the upper and lower sheets of F. We call these *heeling cycles* if the cusp branches upwards and *capsizing angle* if the cusp branches downwards.

Catastrophe Model for Rolling

An elementary catastrophe model is a parametrized system of gradient like differential equations specified by 4 conditions:

(i) a parameter space C
(ii) a state space X,
(iii) an energy function $H : C \times X \rightarrow R$, and
(iv) a dynamic D on X, parametrized by C, that locally minimizes H.

The function H determines the equation manifold $E \subset C \times X$, by the evaluation $\nabla \times H = O$. The catastrophe map $X : E \rightarrow C$ is induced by projection. The bifurcation set is the image in C of the singularities of X.

CATASTROPHE OF MODEL

(1) Define the parameter space to be the plane C containing one 2-dimensional ship. The parameters $G \epsilon C$ is the position of, e.g. the ship.

(2) Define the configuration space to be the unit circle S. The configuration of the ship is uniquely determined by the angle $\theta \epsilon S$. Define the *state space:* $X = T * S$, to be the cotangent bundle of S. The state of the ship is given by $(\theta, \omega) \epsilon T * S$ where ω is the angular momentum. Let

W = weight of the ship

I = moment of inertia of the ship and entrained water

$h = h\,(G, \theta)$ = height of G above B_θ, at an angle $\theta = ZB_\theta$

Lemma. *The P.E. of the system is*

$$P = P\,(G, O) = Wh$$

K.E. of the system is

$$K = K\,(\omega) = \frac{\omega^2}{2I}$$

Define the energy function of the model to be the Hamiltonian H : $C \times T * S \rightarrow R$ given by

$$H = P + K = Wh + \frac{\omega^2}{2I}$$

Lemma $\dfrac{\partial h}{\partial \theta} = l$

The buoyancy locus has coordinate θ and is contained in the space C. Therefore, the normal bundle NB of B is defined by the geodesic spray is the natural cusp $NB \rightarrow C$ of the normal bundle into the S space induced by the projection $C \times S \rightarrow C$. The image of the singularities of the geodesic spray in the evolute of B, which we have called the metacentric locus $\mathcal{M}$.

Theorem. *The equation surface E is the normal bundle NB, the buoyancy locus.*

The catastrophe map $X: E \rightarrow C$ is the geodesic spray. The bifurcation set is the metacentric locus $\mathcal{M}$. We would now complete the model by defining the dynamic.

Assuming G fixed, the Hamiltonian dynamic on $T{*}S$ is uniquely determined from H by the intrinsic sympletic structure of the cotangent bundle. Explicity the dynamic is given by the Hamiltonian

$$\dot{\theta} = \frac{\partial H}{\partial \omega} = \frac{\omega}{I};\ \dot{\omega} = -\frac{\partial H}{\partial \theta} = -W\frac{\partial H}{\partial \theta} = -Wl$$

Therefore

$$I\ddot{\theta} = \dot{\omega} = -Wl$$

Define the dynamic D of the catastrophe model to be the Hamiltonian dynamic with nonnegative damping. There is no need to be any more specific about the nature of damping other than saying that energy is dissipated, because this assumes that H decreases along the orbits of D. Therefore H is a function for D. Therefore, D locally minimizes H, and depends upon the parameter G, as required now.

The model is now complete.

THREE

Applicable Catastrophe Theory II light caustics

E. C. ZEEMAN

Light caustics are the bright geometric patterns created by the reflected or refracted light. A familiar example is the cusp appearing on a cup of coffee in bright sunlight, caused by the reflection of the sun's rays off the inside of the cup (Fig. 1).

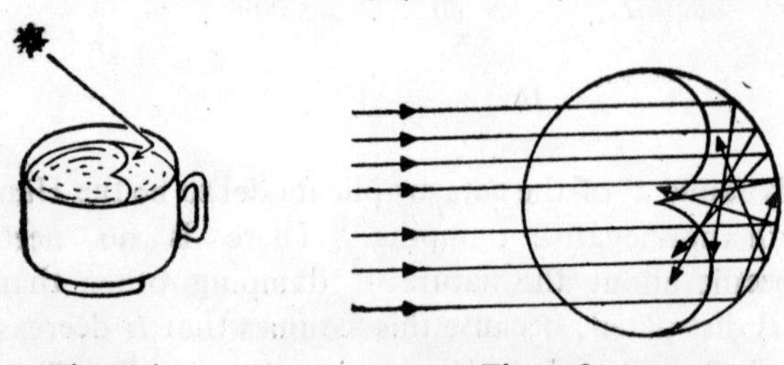

Figure 1 Figure 2

Looking down on the cup from the above Fig. 2 shows that the vertical planes containing the sun's rays after they have been reflected, envelop a vertical surface with cusp shaped horizontal cross-section which is called the caustic. Since the planes all touch the caustic there is a concentration of photons near the caustic (on its convex side). But, of course, these photons are invisible to the eye, because each is travelling along its appointed route. We can only see them if we place a screen in the way where the screen is the surface of the coffee, then the concentration of the photons hitting the screen will be scattered into our eye, causing the cusp-shaped selection of the caustic to appear bright, with a sharp edge on the concave side and a soft edge on the convex side.

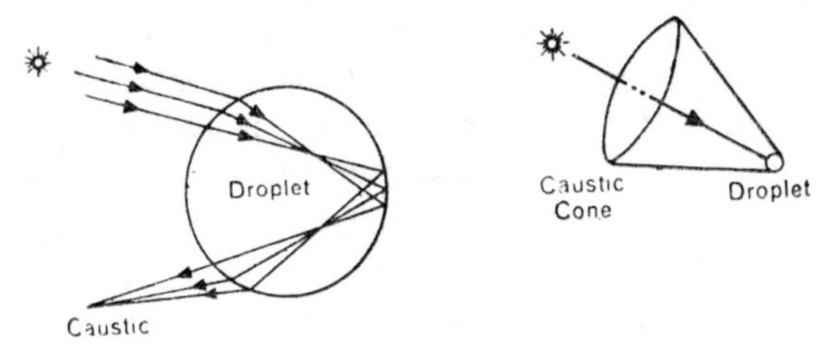

Figure 3. The rainbow is caused by a spectrum of coloured caustic cone produced by each droplet, each cone being a fold catastrophe.

Another familiar caustic is the rainbow (Fig. 3). Here each water droplet refracts and reflects the sun's rays to form a caustic surface approximately core-shaped, with radial angle between 40° and $42\frac{1}{2}$°. The angle depends upon the wavelength, so that, each droplet produces a caxial family of differently coloured caustic cones. Those drops whose caustics happen to meet our eye produce the rainbow.

Catastrophe theory applies to light caustics because light obeys a variational principle: by Fermet's principle the light rays travel along geodesics. Let C be a 3-dimensional neighbourhood of a caustic, and let X be a 2-dimensional surface normal to the incident rays. Each point x in X determines a ray C_x in C, and the union of all these rays, parametrized by X, forms a 3-dimensional manifold.

$$M = \{(c, x) : c \in C_x\} \subset C \times X$$

Since all the rays leave X normally, M is given by $\partial f/\partial x = 0$, where $f(c, x)$ is the geodesic distance from x to c. The intensity of light in a volume element dc under projection $M \to C$. The caustic is where this intensity is the greatest, namely, on the bifurcation set. Therefore, stable caustics are stable bifurcation sets, in other words elementary catastrophes. Therefore the classification theorem gives the new result in geometrical optics: the only stable singularities a caustic can have, besides, cusped edges are the three types of singular points, the swallow tail and the elliptic and hyperbolic umbilics. This discovery of Thom's about caustics was one of the factors that stimulated him to develop catastrophe theory.

Due to Michael Berry photographs of the refraction of a laser beam in frosted glass can be obtained. Since the frosting was periodic in two directions the global picture is the projection of a torus in the plane (each little square of frosting is projected onto the whole picture). Each corner is a section of a hyperbolic umbilic. To analyse the fine structure of interference patterns merging in Airy and Pearcey patterns on the caustic, it is necessary to pass from geometrical optics to the more delicate wave optics. Berry uses the frosted glass as an optical analogue to study the scattering of beams of particles from a solid surface.

FOUR

Applicable Catastrophe Theory III the analysis of some discontinuous processes

J.Q. SMITH
P.J. HARRISON
E.C. ZEEMAN

INTRODUCTION

This paper was motivated by the wish to model many of the situations which lead to sudden change in decision on the part of policy makers and individuals, polarity of opinions and group conflict. Sudden changes in decision may arise in a number of ways. The environment of the decision maker may change slowly, gradually altering his perception of the situation until his current policy is no longer appropriate and a major change is forced. Alternatively such a gradual movement in the environment may cause a sudden qualitative change in the decision makers' perception, as arises from 'contradiction' situations typified by commodity market crashes, confidence tricks and expedient decisions. Polarity in opinion and group conflict obviously arise when one group of people possesses either totally different information or totally different utilities from the other group. However, what is not so well recognized is the fact that such opposition can arise naturally in situations in which the differences in utility and opinion are marginal. It is very important to recognize such situations since the critical change can surprise with its suddenness, arising very quickly from seemingly insignificant environmental changes. Furthermore, if the situation is immediately recognized decision makers may well be able to rectify the situation temporarily by taking relatively small measures. The kind of applications we have in mind range from brand switching (Chidley, 1976) and voting behaviour through to industrial conflicts, prison riots and war (Zeeman, 1977).

In this paper a Bayesian approach is adopted. Individuals or decision centres receive information which, like their utilities, evolves over time. Formally these are parameterized and may be expressed in state-space form with updating using Bayesian forecasting (Harrison and Stevens, 1976). The expected loss associated with any particular person or decision centre is then calculated in the usual way. We are particularly interested in those situations in which the expected loss is a smooth function of the decision and the parameters. In this case we treat the expected loss as a potential function in which the set of Bayes decisions is a subset of its stationary values. Thus we see the relevance of Catastrophe Theory (Thom, 1972) which is concerned with the classification of such potential functions and which gives geometric insights into the nature of the set of Bayes decisions. Where only one Bayes decision corresponds to the environmental parameters, the implemented decision changes slowly with the parameters. However, where there is more than one corresponding decision a selection rule determines which decision is taken and thus an infinitesimal change in the parameters can result in a major or 'catastrophic' change in decision.

Bayesian Decision Theory is introduced in section 1 of the paper with particular emphasis on the wisdom of using bounded loss functions. Section 2 then looks at cases in which a decision maker's information is characterized by a continuous unimodal probability density function and in which the loss function is symmetric with just one minimum. Intuitively it was imagined that this would not produce catastrophic decision changes in response to incremental parameter changes but surprisingly this is not the case. In fact with well used inverse polynomial tailed distributions such as the log-normal, the inverted gamma and most F distributions catastrophic responses in decision arise quite naturally with the simplest of loss functions. Applications are given including one based upon a Pareto distribution because of the resultant dichotomous Bayes decisions.

In section 3 the very innocent looking case of a normal probability density and its conjugate loss function is examined. It may be thought that this is a very stable situation in which the Bayes decision varies smoothly with the natural parameters. However, this is only true in what are often naive models. For example, in practice it usually happens that the uncertainty of outcome and the penalty cost of a mistake depend upon the magnitude of change in decision. For example, in a given capital investment situation both the uncertainty and the penalty of forecast errors with resultant unit produce price will usually depend on the additional sanctioned production capacity. When this dependence is investigated it is found that the phenomena of delayed, discrete and conservative actions are naturally captured.

Section 4 gives a brief introduction to catastrophe theory and its uses, introducing the two most elementary catastrophe singularities, namely, the fold and the cusp. Section 5 deals with multimodal utility functions and/or multimodal probability density functions, which arise, for example, in the multiprocess forecasting models of Harrison and Stevens (1976). Where the resulting expected loss function has two minima, Theorem 5.1 shows the relationship with the cusp catastrophe (Smith, 1979). These results are useful for modelling conflict between two groups which may simply be two competing composite hypotheses as in Example 5.1 or for modelling the different behaviour of groups of people as is illustrated in the drunken driver Example 5.2.

1. Bayesian Decision Theory

A Bayes decision δ^* about a future outcome $\theta \in \Theta$ of a particular process is an argument δ giving the infimum value to an expected loss function $E(\delta)$ where δ represents an element in the class D of possible decisions. $E(\delta)$ is defined by the relation

$$E_{u,v}(\delta) = \int \Theta \, L_v(\delta, \theta) \, dF_u(\theta \mid \delta) \qquad \delta \in D$$

where $L_v(\delta, \theta)$ is a function representing the loss when we choose decision $\delta \in D$ and the true outcome is θ and where the distribution function $F_u(\theta|\delta)$ represents our beliefs about the outcome given that decision δ is employed. Vectors $\mathbf{u} \in \mathbf{U}$ and $\mathbf{v} \in \mathbf{V}$ parameterize F and L respectively and hence the expected loss $E(\delta)$. As the environment changes and we collect more information so $\mathbf{u} \in \mathbf{U}$ and $\mathbf{v} \in \mathbf{V}$ change in sympathy. Thus the expected loss function $E_{u,v}(\delta)$ and the corresponding Bayes decision $\delta^*(u, v)$ can be thought of as functions of these parameters.

Two assumptions are usually made about L_v and F_u. The first is that L_v is convex in $(\delta - \theta)$. For most problems this kind of loss function is both unrealistic and theoretically dubious (see Kadane and Chuang, 1978). The second assumption about $E(\delta)$ is that the distribution function $F_u(\theta)$ of θ does not depend on δ. In section 3 of this paper we will illustrate one of the many types of situations when this is in fact not the case.

The combination of these two assumptions will nearly always imply that a Bayes decision $\delta^* \in \mathcal{R}$ will vary continuously under smooth changes in the parameters (u, v) of F_u and L_v. However, if one of the above mentioned assumptions is not appropriate then there are often values of **u** and **v** such that $E_{u,v}(\delta)$ has more than one local minimum. This case, as $E(\delta)$ changes, its infimum may jump between these minima. Then the corresponding Bayes decision $\delta^*(u, v)$ can be a very discontinuous function of **u** and **v**.

If we are not to use convex loss functions then we need some convenient families of non-convex loss functions to illustrate certain problems. One such family is given by Lindley (1976) who specifies loss functions *conjugate* to certain distributions. For example, the conjugate loss function $L_v(\delta, \theta)$ to the normal distribution with mean μ and variance V satisfies

$$L_v(\delta, \theta) = h(1 - \exp\{-\tfrac{1}{2} k^{-1} (\theta - \delta)^2\})$$

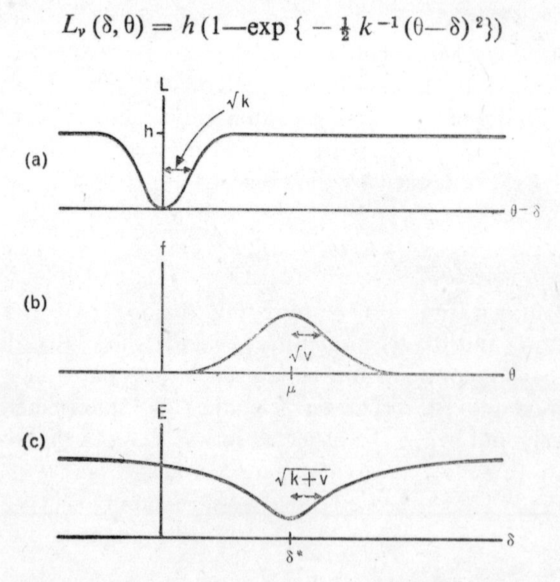

Figure 1.1. (a) Loss, (b) density and (c) expected loss.

where $\mathbf{u} = (\mu, V), \mathbf{v} = (, hk)$ and $V, h, k > 0$. The corresponding expected loss function is given by

$$E_{u,v}(\delta) = h \left[1 - \left(\frac{k}{k+V}\right)^{\frac{1}{2}} \exp\{-\tfrac{1}{2}(k+V)^{-1}(\delta - \mu)^2\}\right]$$

Note that as $k \to \infty$ the Bayes decision associated with L_v tends to the posterior mean and as $k \to 0$ the Bayes decision tends to the posterior mode. This combination of loss function and distribution will be used frequently in examples in this paper.

If h and V are constant then $E_{u,v}(\delta)$ has only one local minimum, at $\delta = \mu$ as shown in Fig. 1.1c. However we shall be particularly concerned below with examples where k and V depend on δ and in these cases E may, surprisingly, turn out to have two local minima, as shown in Fig. 1.2.

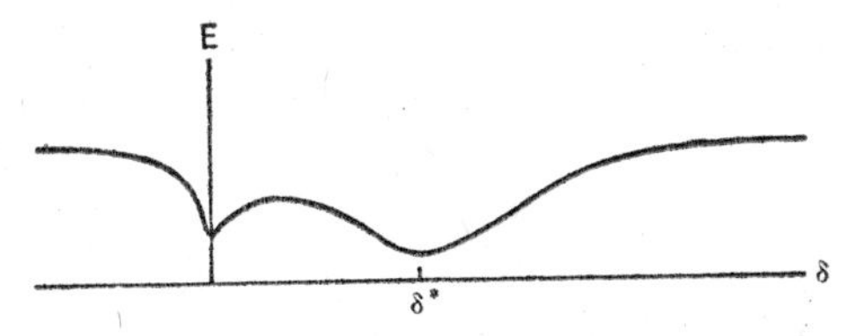

Figure 1.2. Bimodal expected loss, if k increases with $|\delta|$.

2. Expected Loss Functions from Skew Distributions

In this section we shall consider some of the conditions on unimodal probability distributions and loss functions that give rise to more than one minimum of the corresponding expected loss function.

Let $\mathcal{L}$ be the class of loss functions $L_v(\theta - \delta)$ that are bounded monotonic increasing functions of $|\theta - \delta|$ and parametrized by $\mathbf{v} \in \mathbf{V}$, where $\theta \in \mathcal{R}$ is the outcome, and δ is the decision about θ. Let $\mathcal{F}$ be the class of distribution functions $F(\theta)$ of outcome θ which are twice differentiable at all points θ satisfying $f(\theta) > 0$ where $f(\theta)$ is the density of $F(\theta)$. The expected loss function is given by

$$E_v(\delta) = \int_{\mathcal{R}} L_v(\theta - \delta) f(\theta)\, d\theta$$

It would be tempting to assume that the above type of symmetric loss function used on a unimodal density of the kind mentioned above would give rise to an expected loss function with one minimum. Ibragimov's (1956) solution to an analogous problem involving the convolution to two unimodal distributions has been adapted by Smith (1978) to give the following theorem exhibiting sufficient conditions on the density for this to be so.

Theorem 2.1. *Let the distribution function $F(\theta) \in \mathcal{F}$ and have density $f(\theta)$. Let $\tau(\theta) = d(\ln f(\theta))/d\theta$ be strictly decreasing on B_F where B_F is ths set of all $\delta \in \mathcal{R}$ such that $E_{\mathbf{v}}(\delta)$ has a Bayes decision at δ for some $\mathbf{v} \in \mathbf{V}$. Suppose, in addition, that*

$$\tau(\theta) > \tau(d_1) \text{ for } \theta < d_1$$
$$\tau(\theta) < \tau(d_2) \text{ for } \theta > d_2$$

where $[d_1, d_2]$ is the smallest closed interval containing B_F. Then for any loss function $L_{\mathbf{v}} \in \mathcal{L}$, $\mathbf{v} \in \mathbf{V}$ the expected loss function $E_{\mathbf{v}}(\delta)$ corresponding to $L_{\mathbf{v}}$ and F has exactly one minimum.

Examples of distributions satisfying the conditions of the above theorem for all families of loss functions contained in $\mathcal{L}$ include all unimodal distributions with symmetric densities, and the Gamma and Beta distributions. If every member of our family of distribution functions $F_u(\theta)$ $\mathbf{u} \in \mathbf{U}$ satisfies the conditions of this theorem and $L_{\mathbf{v}} \in \mathcal{L}$ for all $\mathbf{v} \in \mathbf{V}$ we can now conclude that no dramatic discontinuous changes in the Bayes decision of δ^* will occur under smooth changes in $\mathbf{u}$ and $\mathbf{v}$.

Although many standard distributions satisfy the requirements of this theorem the conditions it imposes on the density $f_u(\theta)$, $\mathbf{u} \in \mathbf{U}$ are in fact very strong. Smith (1978) has proved some useful partial converses and an important corollary to one of these is the following theorem.

Theorem 2.2 *If the distribution $F(\theta) \in \mathcal{F}$ has the support of its density $f(\theta)$ equal to $[0, \infty)$, and if corresponding function $\tau(\theta) = d(\ln f(\theta))/d\theta \to 0$ as $\theta \to \infty$, then there exists a loss function $L \in \mathcal{L}$ such that the expected loss function corresponding to $F(\theta)$ and $L(\theta - \delta)$ has at least two minima,*

It is easily checked that every distribution in the lognormal, inverted gamma families and most distributions in the F family satisfy the conditions of the theorem, all of which commonly occur as posterior distributions. Furthermore the set of loss functions in $\mathcal{L}$ which cause the expected loss function to have two local minima is large. Perhaps the simplest such loss function is the double step loss (Fig. 2.1) $L_a(\theta - \delta) = S^{\alpha}_{b,c}(\theta - \delta)$ where

$$S^{\alpha}_{b,c}(\theta-\delta)=\begin{cases}0, \text{ for } |\theta-\delta|<b \\ \alpha, \text{ for } b\leqslant|\theta-\delta|<c \\ 1, \text{ for } |\theta-\delta|\geqslant c\end{cases} \quad (2.1)$$

and where b and c are chosen depending on our family $F_u(\theta)\mathbf{u}\in\mathbf{U}$.

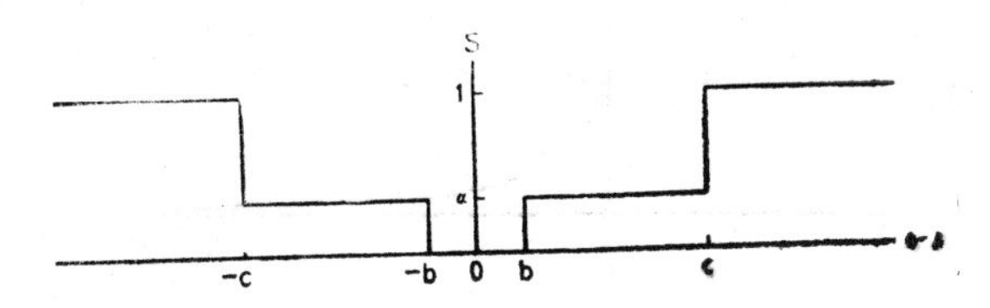

Figure 2.1. Double step loss

To illustrate vividly the splitting of the set of all Bayes decisions into separate interval subsets we look at a case based on the following theorem. Here the set of all Bayes decision corresponding to different values of $\mathbf{u}\in\mathbf{U}$ and $\mathbf{v}\in\mathbf{V}$ consists of just two distinct points.

Theorem 2.3. *Let $F_u(\theta)$ be a distribution function of θ parametrized by* **u**. *Suppose its density $f_u(\theta)$ is differentiable and strictly decreasing on* $(0, \infty)$ *and zero elsewhere. Let $\tau(\theta)=d(\ln f(\theta)d\theta$ be strictly increasing on* $(0, 2c)$ *and $L_\alpha(\theta-\delta)=S^{\alpha}_{b,c}(\theta-\delta)$ defined above. Then the Bayes decision θ^* is given by the formula*

$$\delta^*=\begin{cases}c \text{ if } 0\leqslant\alpha<\alpha^* \\ b \text{ if } \alpha^*<\alpha\leqslant 1\end{cases}$$

where $$\alpha^*=\frac{F_u(2c)-F_u(b+c)}{F_u(2)+F_u(2b)+F_u(c-b)-2F_u(b+c)}$$

When $\alpha=\alpha^*$ the decision is ambiguous since the two local minima at b and c are at the same level.

Example. Let $L_\alpha(\theta-\delta)=S^{\alpha}_{b,c}(\theta-\delta)$ defined above and

$$F(0)=\begin{cases}0 & \theta\leqslant 0 \\ 1-(1+\theta)^{-k} & 0<\theta<\infty\end{cases} \qquad \text{where } k>0$$

which is a Pareto distribution. It is easily checked that the conditions

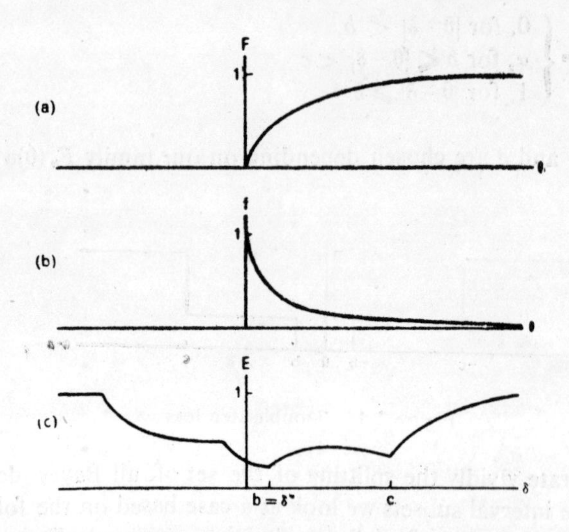

Figure 2.2. (a) Pareto distribution, (b) density, (c) expected loss (drawn for $k=1$, $b=0.05$, $c=3$, $\alpha=0.25$, $\alpha^*=0.15$)

of the theorem are satisfied. In this case therefore we can exhibit two different types of bifurcating behaviour.

Case 1: Suppose we have a population of individuals each of whom is characterized by a parameter vector $(v, u) = (\alpha, k)$ with common values of b and c but differing (α, k)'s. Then given that each individual adopts a Bayesian decision the population polarizes into a group G_1 whose decision $\delta^* = b$ and a group G_2 whose decision $\delta^* = c$ depending upon whether α is greater or less than $\alpha^*(k)$.

Case 2: Consider a single individual whose information about θ occurs sequentially so that the value k varies with time. The information can be summarized in this case by parameters α and k. Even if α, and hence this loss function, remains constant over time, the variation in k alters the value of α^* and as the sign of $\alpha - \alpha^*$ alters so the individual's decision jumps between b and c.

In quality control it is not unusual to encounter loss functions which are close to double step. For example, suppose that the target quality of batch product is $\mu=10$ units with a tolerance of ± 1 unit on blends. Batches within (9, 11) may prove of little problem in blending, batches in the regions (8, 9) and (11, 12) may be blended successfully at a cost of special treatment, handling and temporary storages, and batches outside (8, 12) may need complete reprocessing.

Denoting Q as the batch quality, the loss function is thus approximately given by

$$L_\alpha(Q) = \begin{cases} 0 & |Q-10| \leqslant 1 \\ \alpha & 1< |Q-10| \leqslant 2 \\ 1 & 2< |Q-10| \end{cases}$$

If the batch quality has a distribution satisfying the conditions of either Theorem 2 or Theorem 3, then slight variations in the parameters of these distributions may lead to major jumps in control action. For example for computational case suppose that the "inherent" quality of a batch follows a Pareto distribution

$$F(Q)= \begin{cases} 0 & Q \leqslant 0 \\ 1-(Q+1)^{-k} & 0 < Q \end{cases}$$

and that, subject to control action equivalent to q, the quality is shifted so that $F_q(Q)=F(Q-q)$. Then the expected loss under q is

$$\begin{aligned} \int L_\alpha(\mu - Q)\, dF_q(Q) &= \int L_\alpha(\mu - q - \theta)\, dF(\theta) \qquad \text{putting } Q = q - \theta \\ &= \int L_\alpha(\delta - \theta)\, dF(\theta) \qquad \text{putting } \delta = \mu - q \\ &= E_{\alpha,q,k}(\delta) \end{aligned}$$

which takes its minimum value when $\delta = \begin{cases} 1 & \alpha > \alpha^* \\ 2 & \alpha < \alpha^* \end{cases}$. Taking $k = 1$, for example, it is easily found that $\alpha^* = 0.107$ so that if $\alpha > 0.107$ the Bayes decision is $\delta^* = 1$ (or equivalently $q^* = 9$) but if $\alpha < 0.107$ then it is $\delta^* = 2$ or $q^* = 8$. In practice as the production, which may be parametrized by k, varies, so α^* varies. Thus an optimal control action may undergo sudden jump changes. For example, if the quality of production improves k will increase so α^* will decrease; and if it decreases past α the optimal control q^* will switch from 8 to 9. Meanwhile if the cost of special treatment decreases, then α will decrease; and if it decreases past α^* then optimal control will switch from 9 to 8.

Finally it should be noted that we could, for example, use an inverted gamma distribution for the quality of the batch instead of the Pareto to obtain similar jumps in optimal behaviour. This example would have the advantage of keeping the expected loss smooth but is computationally rather messy.

3. Delayed Action and Adaption

In this section we look at one way of modelling delayed action and adaption. This involves, for each of its possible actions, a decision centre having a unimodal belief distribution $F(\theta)$ about the future outcome and a loss function $L(\delta, \theta)$ relating the utility of outcome and action. In particular we shall consider those situations where not only the location of $L(\delta, \theta)$ and $F(\theta)$ but also their variations depends upon the action. For example, generally the more unfamiliar is an action the more uncertain is the corresponding outcome. Similarly, the greater the effort or cost associated with an action, the greater the need for accurate information. We shall illustrate these principles by an example using a normal belief distribution and conjugate loss function.

3.1. *An Illustration Using the Normal Distribution.* Consider a decision centre concerned about the future outcome θ. Their belief is that without action (i.e. $\delta = 0$) the outcome is described by a Normal distribution with mean c and variance V

$$(\theta \mid \delta = 0) \sim N[c, V]$$

However they believe that their actions will influence the outcome in such a way that for action corresponding to decision $\delta \in \mathcal{R}$

$$(\theta \mid \delta) \sim N[c + \delta, V]$$

The parameters of the model can be interpreted in the following way. The parameter may be thought of as the "natural" system outcome expected by the decision centre if they do not act to change their policy, plans or status quo. The decision δ is measured by the effect that the resulting action has upon the outcome and V describes the uncertainty of belief or the variety of opinion within the decision centre or both.

Given the beliefs described above, the decision centre is assumed to have a goal or most desirable outcome μ for which the associated loss is $L(\theta = \mu) = 0$. For analytic purposes and its reasonable character we shall adopt loss functions belonging to the family conjugate to the normal distribution mentioned in the first section

$$L(\theta) = h[1 - \exp\{-(2k(\delta))^{-1}(\theta - \mu)^2\}]$$

The interpretation of μ is given above. k represents the relative tolerance to differences between θ and μ and in general will depend upon the decision δ. For example if k is very small then any variation between θ and μ will give almost maximum loss whereas if k is very large the difference between θ and μ has to be very large before we approach the maximum possible loss. The parameter h represents the largest loss that could possibly be incurred.

It now follows that the expected loss $E(\delta)$ associated with a decision δ is given by

$$E(\delta) = h\left[1 - \left(\frac{k}{k+v}\right)^{\frac{1}{2}} \exp\{-[2(k+v)]^{-1}(\delta - d)^2\}\right]$$

where $d = \mu - c$ represents the distance of the desired value of θ from the expected value of θ.

Clearly if k and V do not depend on δ then $\delta = d$ is the Bayes decision. Notice that d is the decision taken in the absence of uncertainty in belief. However when we consider practical problems we find that usually at least one of k and V do actually depend on δ.

Example 3.1. Here we give a simplified example to illustrate this theme. In considering a full investment programme a decision centre evaluates extending the production capacity of a product. Let P be the unit price of the product and let the belief of the centre about P, given an action with corresponding decision δ, be such that the logarithm of the outcome price is described by the distribution

$$(\theta \mid \delta) \sim N(c - \delta, V)$$

where $\theta = \ln P$ and where δ represents the logarithm of the proportional increase in productivity.

The location $c - \delta$ reflects the relationship of supply and price, in that, for constant demand, supply and price vary inversely.

We will assume the unit production costs are constant and the target contribution of the investment programme is such that $\theta = \mu$. If the outcome is lower than μ this would indicate unwise over-investment or if the outcome is greater than μ this indicates a missed opportunity resulting from under-investment. Thus for illustrative purposes it is not unreasonable to adopt a conjugate loss function

$$L(\delta, \theta) = h\,[1 - \exp\{-(2k)^{-1}(\theta - \mu)^2\}]$$

However it is clear that, in practice, both the uncertainty associated with the outcome price and the loss function will depend on the size of the action taken. For example if the decision centre invests a few million pounds to increase production dramatically as opposed to a few thousand to increase production marginally, the loss associated with a given error of unit price from the target is very much greater in the former than in the latter case. Similarly the uncertainty associated with the effect on price of a large increase in supply is usually far greater than with a moderate increase.

In studying the delays and sudden changes in decision we will consider a developing environment in which d increases smoothly with time from an initial value of zero. That is, initially the decision centre is happy that their current policy is ideal. However, as the environment evolves with time so d evolves and actions need to be reviewed. Suppose that the loss function parameter k is dependent upon the action according to

$$k = k(\delta) = (\pi + \rho\,|\,\delta\,|)^{-1}, \ \pi, \rho > 0$$

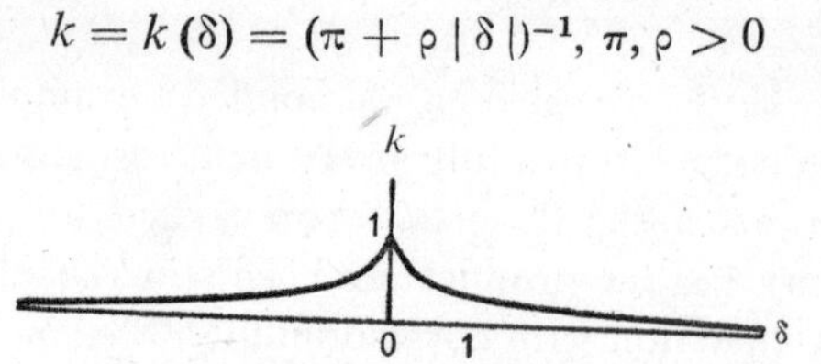

Figure 3.1. k decreases with $|\,\delta\,|$ (drawn for $\pi = 1, \rho = 2$).

but that V is constant. If we measure δ in units of $\sqrt{V}$ and define $\gamma = \rho(1 + \pi)^{-1}$ and $\omega = \pi(1 + \pi)^{-1}$ so that $0 < \omega < 1$, then minimizing the expected loss $E(\delta)$ is equivalent to minimizing the function

$$S(|\,\delta\,|) = \ln(1 + \gamma\,|\,\delta\,|) + \frac{(\omega + \gamma\,|\,\delta\,|)(\delta - d)^2}{1 + \gamma\,|\,\delta\,|}$$

We assume that $d = 0$ initially and subsequently will always remain greater than or equal to zero. It is then clear that we need only be concerned with Bayes actions associated with decisions $\delta \geqslant 0$. Firstly we note that the derivative of S with respect to δ, $S'(\delta)$ satisfies

$$S'(\delta)|_{0+} = \gamma - 2\omega d + \gamma d^2(1-\omega)$$

which shows that for $0 < d$ sufficiently small, $\delta = 0$ gives a local minimum of $E(\delta)$. In fact if $\omega^2 < \gamma^2(1-\omega)$ (i.e. $\rho > \pi\sqrt{1+\pi}$) then $\delta = 0$ is a local minimum for all $d \in \mathcal{R}$; we can see from this that ρ represents a penalty for bold decisions. If this is not the case then $\delta = 0$ is a local minimum if and only if

$$0 \leqslant d < \frac{\omega - (\omega^2 - \gamma^2(1-\omega))^{1/2}}{\gamma(1-\omega)} \quad \text{or } d > \frac{\omega + (\omega^2 - \gamma^2(1-\omega))^{1/2}}{\gamma(1-\omega)}$$

Next it is easily found that if $\delta > 0$ then $S'(\delta) = 0$ is a cubic equation whose positive real roots all lie in $(0, d]$.

Figure 3.2 shows the section of the graph $S'(\delta) = 0$ in units of $\sqrt{V}$ for $\pi = 1$, $\rho = 2$ and $d > 0$. It is seen that the decision $\delta = 0$ remains the only minimum of $E(\delta)$ until $d = 1.04$ when a competing minimum

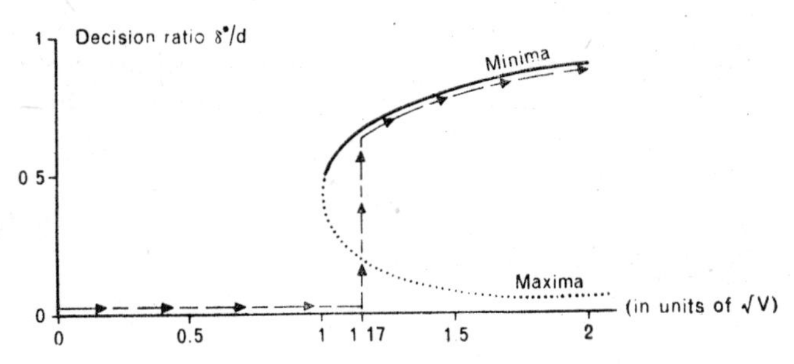

Figure 3.2. Expected loss $E(\delta)$ for $d = 0.5$, 1.17 (critical), 2.

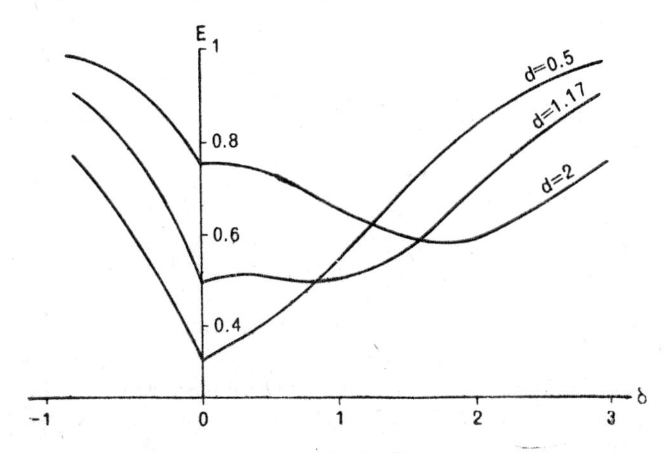

Figure 3.3. Delay in adaption due to changes in k.

arises at $\delta = 0.41 = 0.40d$. However $\delta^* = 0$ remains the Bayes decision until $d = 1.17$ when there is a sudden jump in Bayes decision δ^* between $\delta^* = 0$ and $\delta^* = 0.75 = 0.64\,d$. The Bayes decision as d increases is shown by the dashed path of Fig. 3.2. The corresponding graphs of $E(\delta)$ for three particular values of d (including the critical value 1.17) are shown in Fig. 3.3, with dots indicating the Bayes decision.

This one example illustrates that where the cost of an error in taking a course of action depends upon the departure of that action from the status quo it is to be expected that decision centres will delay revising their actions until the pressure between the perceived demands of the environment d and the current decision centre action δ^* reaches or exceeds a given threshold. Furthermore, the extent of the delay is clearly larger the more uncertain the outcome (i.e. the greater V).

Where a number of decision centres are considering similar production extensions or investment the population of decision centres may be parametrized. For example, the main parameters might be d and V in which case the pertinent geometry describing the way that the Bayes decisions vary with the parameters is equivalent to that of a canonical cusp catastrophe truncated at zero. The cusp catastrophe is described in Section 4.

From the geometric point of view the fact that $E(\delta)$ is not smooth at $\delta = 0$ in this example makes it rather inelegant. This lack of smoothness can be removed however by using a different (but similar) function $k(\delta)$ in our loss function, namely

$$k(\delta) = (\pi^2 + \rho^2\,\delta^2)^{-1/2}$$

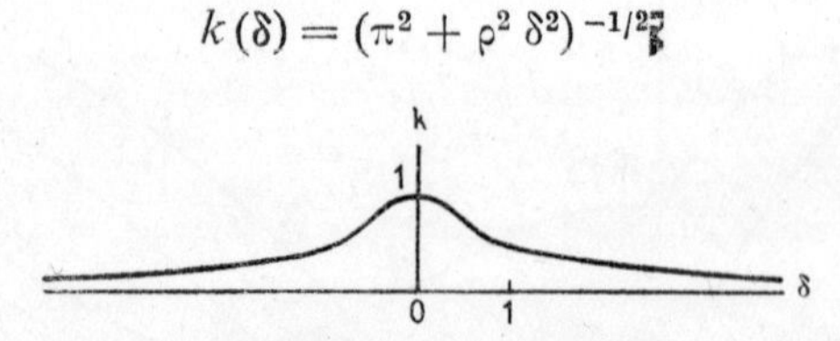

Figure 3.4 Smooth k (drawn for $\pi = 1$, $\rho = 2$).

This time we obtain a full cusp catastrophe in $E(\delta)$. This is illustrated by Figs. 3.5 and 3.6.

We now examine the effects of varying uncertainty on the action of a decision centre. Retaining the normal belief and conjugate loss

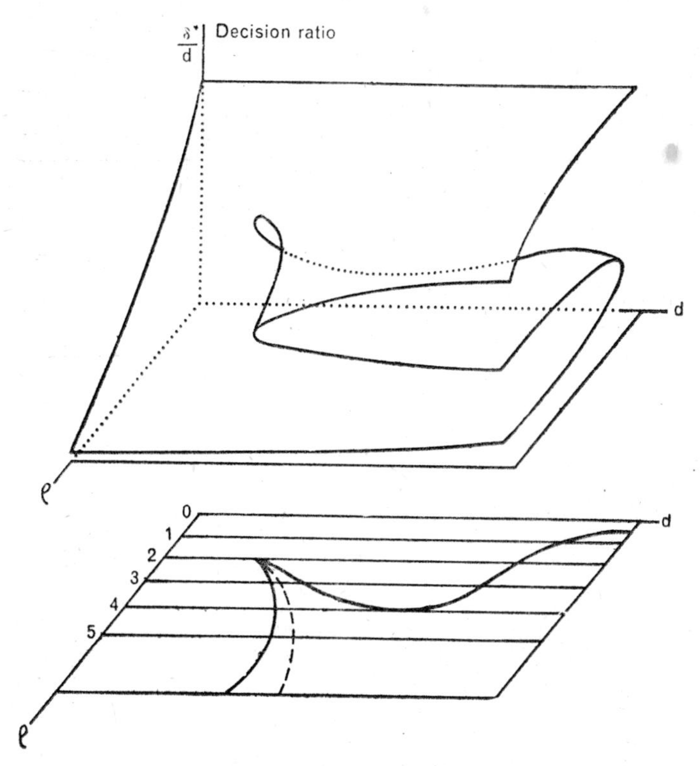

Figure 3.5 Sketch of surface $S' = 0$, as a graph of the decision ratio δ^*/d over the parameters ρ, d. Whenever the parameter vector (ρ, d) crosses the dotted line, the Bayes decision jumps dramatically from one value to another.

the uncertanity V is treated as a function of the action. In practice the variance will generally be an increasing function of the degree of change involved in taking a new decision. Consider first the case in which

$$V = V(\delta) = \alpha + \beta |\delta| \qquad \alpha, \beta > 0$$

and in which $d > 0$. It is then easily shown that the Bayes decision $\delta^* \in [0,d]$ and that minimizing the expected loss $E(\delta)$ is equivalent to minimizing the function

$$S(\delta) = \ln(k + V(\delta)) + (\delta - d)^2 (k + V(\delta))^{-1}$$

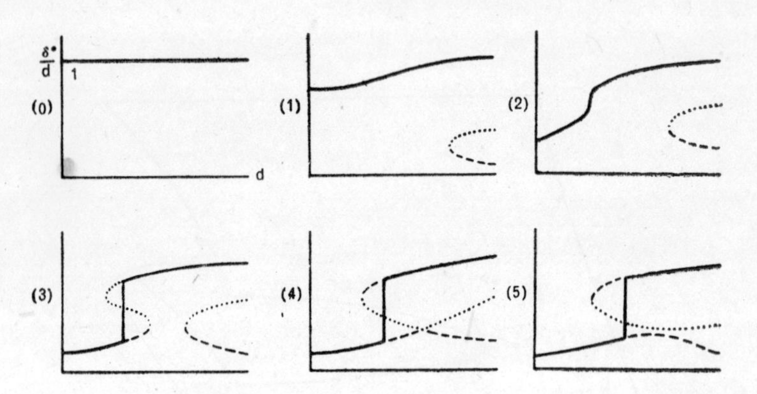

Figure 3.6 Sections ρ = constant of the surface $S' = 0$ over the six lines marked 0–5 in Figure 3.5.
Global minima (= Bayes decision)———
Local minima-·-·-
Local maxima . . .

Measuring δ in units of $\sqrt{k + V(0)} = \sqrt{k + \alpha}$ and defining $\gamma = \beta(k+\alpha)^{-1/2}$ then since $S'(\delta) = 0$ for $\epsilon > 0$ is a quadratic, and examining separately $\delta = 0$, it is found that provided $d > 0$ the Bayes decision is

$$\delta^* = \begin{cases} 0 & \text{for } 0 < d < ((1+\gamma^2)^{1/2} - 1)\gamma^{-1} \\ [(4(1+\gamma d)^2 + \gamma^4)^{1/2} - (2+\gamma^2)]/2\gamma & \text{otherwise} \end{cases}$$

and as $d \to 0$

$$\lim (d - \delta) = \tfrac{1}{2}\gamma, \text{ and}$$

$$\lim \frac{\delta^*}{d} = 1.$$

The rate of change of δ^* with d is thus zero up to a threshold value indicating that the decision centre does not change its decision until d exceeds a given value. At this point the transition between non-adaption takes place, the rate of adaption increasing with d with asymptote 1. Figure 3.7 shows the adaption rate for various values of γ.

It is interesting to note that if we generalize the variance relation to

$$V(\delta) = \alpha + \beta\delta^{\rho} \qquad \rho > 0$$

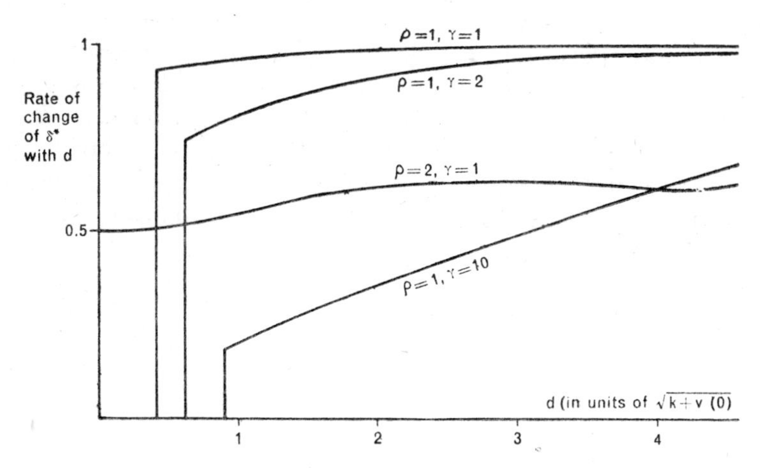

Figure 3.7 Delay in adapation due to changes in V.

where $\rho \in (0, 1]$ the qualitative behaviour of the Bayes decision δ^* is the same as that above, i.e.

$$\lim_{d \to \infty} \frac{\delta^*}{d} = 1$$

This is not the case if $\rho \in (1, \infty)$. For example if $\rho = 2$ and $\gamma = 1$ then $\delta^*/d \in [0.5, 0.62]$ and increases with d to a limit of 0.62. So in this case one never dares a decision within 38% of that which one would take if the uncertainty was constant or if there was no uncertainty.

4. Catastrophe Theory and Its Uses

In the previous sections of this paper we have studied the behaviour of the Bayes decision $\delta^*(u, v)$ corresponding to the expected loss function $E_{u,v}(\delta)$ for a few specific problems. Catastrophe theory geometrically classifies smooth functions like $E_{u,v}(\delta)$ in terms of their parameters (u, v). Before we try and prove any *general* theorems about discontinuous behaviour it is, therefore, important to be aware of this classification.

Let M denote the surface $d(E_{u,v}(\delta))/d\delta = 0$ of stationary values of $E_{u,v}(\delta)$ over the space $D \times U \times V$. Then if there is only ever one point δ on M for each value of (u, v) we know that $E_{u,v}(\delta)$ has just one stationary point (usually a minimum) regardless of (u, v). It will follow that if (u, v) changes smoothly then so will the Bayes decision $\delta^*(u, v)$ corresponding to $E_{u,v}(\delta)$. If, however, the surface M folds over on itself somewhere so that there are values of (u, v) in a region $U^* \times V^*$ (say) such that there is more than one point $\delta(u, v)$ on M then there may be more than one minimum of $E_{u,v}(\delta)$ for $(u, v) \in U^* \times V^*$. In this case the Bayes decision δ^* may be a discontinuous function of (u, v) and jumps in δ^* from one local minimum to another may occur as (u, v) smoothly evolves.

For M to fold over on itself in this way there needs to be a non-empty set of points (u, v) such that

$$\frac{d(E_{u,v}(\delta))}{d\delta} = \frac{d^2(E_{u,v}(\delta))}{d\delta^2} = 0$$

This set of points in $U \times V$ is called the *bifurcation set* B and $C = U \times V$ is called the *control or parameter* space. Catastrophe theory classifies $E_c(\delta)$, $c \in C$, in terms of the appearance of B (distinct topological types). If C is a vector space of dimension $\leqslant 2$ and $E_c(\delta)$ is a fixed stable family of expected loss functions then, in a sufficiently small open set $C^* \subset C$, $C^* \cap B$ is always one of the following three sets: (i) the empty set, (ii) a single smooth curve and (iii) a smooth curve with a single cusp in it.

If case (iii) occurs we can show that the cusp in B will look like the cusp we get in the bifurcation set of a quartic expected loss function. The bifurcation set corresponding to this quartic is called the *canonical cusp catastrophe.* Locally, where C has dimension no greater than two, this is the most complex geometric form for B provided that the smooth function $E_c(\delta)$ is stable, which almost all are.

The Canancial Cusp Catastrophe

$$D = \mathcal{R} \qquad C = \mathcal{R} \times \mathcal{R} \qquad E_c(\delta) = \delta^4/4 - b\,\delta^2/2 - a\,\delta$$

$$c = (a, b)$$

M is given by $\delta^3 - b\,\delta - a = 0$.

The canoncical cusp is illustrated in Fig. 4.1. The cusp in line B is clearly visible and points in a certain direction. We now choose a coordinate system on the control space $C = U \times V$ such that the

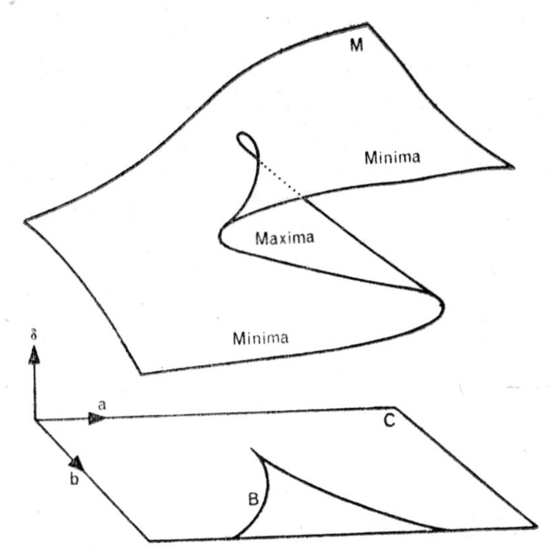

Figure 4.1 The cusp catastrophe.

direction of the cusp is parallel to one of the axes. This axis is called the *splitting factor b*. By convention we let the cusp point have one of its coordinate $b = 0$ and let the direction of the b axis be into the cusp as in Fig. 4.1. Any axis a transversal to b is called the *normal factor*.

Considering Fig. 4.1 it is obvious that if $b > 0$ then there exist values of the normal factor a for which the quartic expected loss function has two minima and one maximum. Therefore, as the normal factor a smoothly increases from a large negative value and $b > 0$ remains constant, the Bayes decision δ^* with respect to this expected loss will undergo a discontinuous trajectory. This is not only true for the quartic but also true for any expected loss function with a single cusp in it. On the other hand, when the splitting factor $b < 0$ and we vary the normal factor a smoothly, the Bayes decision δ^* will change smoothly in sympathy.

Thus if we can prove algebraically that a particular expected loss function $E_c(\delta)$ has only one cusp in its bifurcation set B then we know that we can transform the coordinates of C so that the statements above are also true for δ^* associated with $E_c(\delta)$. Of course, the use of this classification will depend on how often we can expect to encounter expected loss functions with a single cusp. We have already seen one example in Section 3 of a cusp catastrophe and the cusp is illustrated in Fig. 3.5. Clearly we could paraphrase the type of discontinuous behaviour captured by this family of expected loss functions by describing it as giving rise to a single cusp catastrophe with its cusp lying in a certain given position and its splitting factor pointing in a certain given direction. This description pertinently summarizes how the Bayes decision $\delta^*(u, v)$ will vary under smooth changes in the parameters (u, v).

In the next section a theorem is given which connects a general class of naturally occurring families of expected loss functions to the cusp catastrophe. We can then use dynamic interpretations arising from this geometrical classification to describe qualitatively how the Bayes decision will behave under changing parameter values.

Finally it will be shown how, *under one descriptive heading*, the cusp catastrophe links three simple, but at first sight dissimilar, decision processes together with the example in Section 3.

5. A Cusp Catastrophe in Bayesian Decision Theory

A key use of the geometry of the cusp catastrophe in Bayesian estimation theory is given in the following theorem. Given a function E of s, let E', E'', E'' denote respectively the first, second and third derivatives of E with respect to s. We say $E(s)$ is of type T if it is C^∞, symmetric, strictly increasing in $|s|$, with $\lim_{s\to\infty} E(s) = 1$ and satisfyings the three conditions

(i) E'' has one zero in $(0, \infty)$ at η, say
(ii) E''' has one zero $(0, \infty)$ at λ, say
(iii) the images of $(0, \eta), (\eta, \lambda)$ under the function E'''/E' have empty intersections.

For example $E(s) = 1 - \exp(-\frac{1}{2}s^2)$ is the type T with $\eta = 1$ and $\lambda = \sqrt{3}$.

Theorem 5.1 *Let E be of type T. Let*

$$E^*(\delta) = \alpha E(\delta + \mu) + (1 - \alpha)E(\delta - \mu)$$

defined over the two dimensional parameter space given by $0 < \alpha < 1$ and $\mu > 0$. Then $E^*(\delta)$ exhibits one unique cusp catastrophe whose coordinates are given by

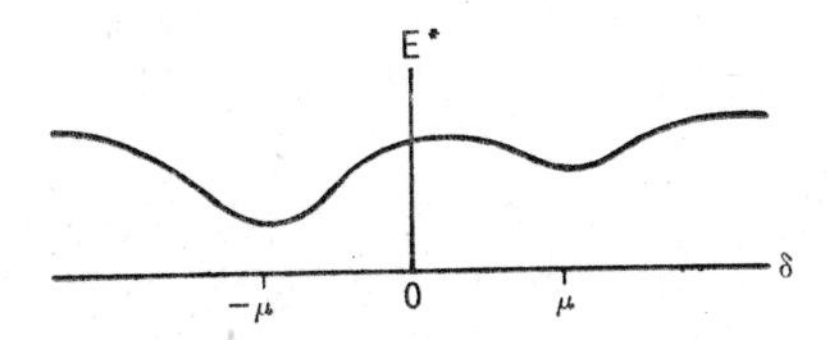

Figure 5.1 Graph of E^*.

$$(\delta, \alpha, \mu) = (0, \tfrac{1}{2}, \eta),$$

wiih normal factor $-\alpha$ *and splitting factor* μ.

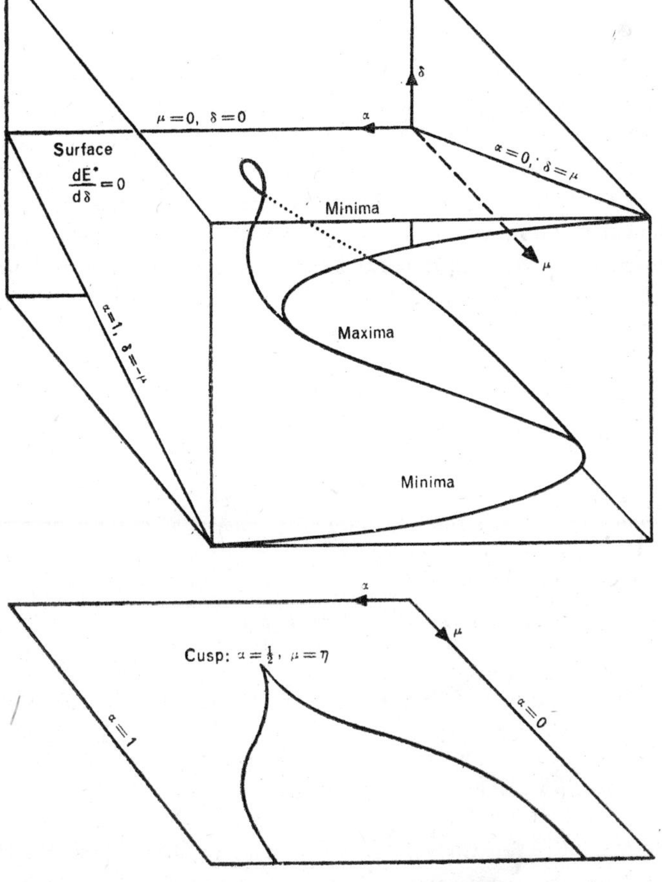

Figure 5.2 E^* exhibits a cusp catastrophe over (α, μ)

The resulting surface of stationary values of E^* is illustrated in Fig. 5.2. The theorem has the following corollary.

Corollary 5.1.1 *If* $J(\delta) = A_1 E(\sigma^{-1}(\delta - m_1)) + A_2 E(\sigma^{-1}(\delta - m_2) + C$ *where* C *may be a function of* $A_1, A_2, \sigma, m_1, m_2$ *but not of* δ, *and where* A_1, A_2: $\sigma > 0$, $m_1 > m_2$ *and* E *is of type* T, *then* $J(\delta)$ *only has cusp points along the coordinates* $(\delta, A_1, A_2, \sigma, m_1, m_2) = (m, A, A, \sigma, m + \sigma\eta, m - \sigma\eta)$, *where* $m = \frac{1}{2}(m_1 + m_2)$ *and* η *is the unique solution in* $(0, \infty)$ *of the equation* $E'' = 0$, *with normal factor* $A_1 - A_2$ *and splitting factor* $m_1 - m_2$.

The proofs of these and related results are given in Smith (1978). Notice that the normal factor is just a function of the mixture variable α in Theorem 5.1 which summarizes all the asymmetry in this particular expected loss function. In contrast the splitting factor is just a function of the distance between the components of this mixture. When the conditions of the theorem relate to a practical situation the results indicate the type of behaviour to be expected from decision-makers. When the two locations in the mixture are close we expect the decision-maker to compromise his decision between these points. On the other hand, when the weights on the locations are of a similar magnitude and the locations far apart then his expected loss function will have two local minima. Thus he will choose his decision close to one or other of the locations and not compromise between them. This kind of qualitative insight helps the construction of many models describing bifurcating behaviour.

Of particular use is the following Corollary.

Corollary 5.1.2 *Let* $E(\delta)$ *represent the expected loss function corresponding to a normal posterior distribution of a parameter* θ *with zero mean and variance* V *together with the conjugate loss function given in Section 1 with parameter* k. *Then the conditions of Theorem 5.1 are met so that the function* $E^*(\delta)$ *defined by*

$$E(\delta) = \alpha E(\delta - \mu_1) + (1 - \alpha) E(\delta - \mu_2) \qquad \begin{array}{c} \mu_1 > \mu_2 \\ 0 < \alpha < 1 \end{array}$$

exhibits cusps only along the coordinates

$$(\delta, \alpha, \mu_1 - \mu_2) = (\tfrac{1}{2}(\mu_1 + \mu_2), \tfrac{1}{2}, 2(V + k)^{1/2})$$

with normal factor α *and spliting factor* $\mu_1 - \mu_2$,

It is interesting to note that in this case it is only possible to have two local minima in the expected loss function with respect to some loss function defined by the parameter k if the variance $\frac{1}{4}(\mu_1 - \mu_2)^2$ "between" the locations of the two expected loss functions is greater than the variance V "within" each component of the mixture.

Here are a few examples of the use of the theorem.

Example 5.1 *A Bayesian Estimation Catastrophe.* Let $X_1, \ldots, X_n$ be independent identically distributed random variables with normal distribution, mean θ and variance τ^2. We wish to estimate θ when τ^2 is known. Our prior beliefs before the experiment began were that there was a probability α and θ was very close to some value μ and a probability $1-\alpha$ that θ could be anywhere. The reason we consider such a prior distribution is that it represents a Bayesian analogue of the beliefs expressed by a sampling theorist when he tests the null hypothesis $\delta = \mu$ against an alternative hypothesis $\theta \neq \mu$.

In this example, a convenient form for such a prior probability density function $g(\theta)$ might be

$$g(\theta \mid \beta, \mu, \sigma) = \alpha(\sigma) f(\theta \mid \mu, \sigma^2) + (1 - \alpha(\sigma)) f(\theta \mid \mu, \sigma^{-2}) \quad (5.1.1)$$

where $0 \leqslant \sigma < 1$, $\mu \in \mathcal{R}$ and $\beta \in \mathcal{R} > 0$ and where

$$f(\theta \mid \mu, \sigma^2) = (2\pi)^{-1/2} \sigma^{-1} \exp\{-\tfrac{1}{2}\sigma^{-2} (\theta - \mu^2\} \quad (5.1.2)$$

and

$$\alpha(\sigma) = \frac{\beta\sigma}{1 + \beta\sigma}$$

This $g(\theta)$ is a mixture of two normal distributions with the same location μ. Characteristically we shall choose σ very small to model the sort of prior distribution described above. After observing $x_1, \ldots, x_n$ the posterior density $g(\theta \mid x)$ is in this case

$$g(\theta \mid \underset{\sim}{x}) = \alpha^* f(\theta \mid \mu_1^*, V_1^*) + (1 - \alpha^*) f(\theta \mid \mu_2^*, V_2^*) \quad (5.1.3)$$

where

$$\mu_1^* = \frac{\sigma^{-2}\mu + n\tau^{-2}\bar{x}}{\sigma^{-2} + n\tau^{-2}}, \quad \mu_2^* = \frac{\sigma^2\mu + n\tau^{-2}\bar{x}}{\sigma^2 + n\tau^{-2}}$$

$$V_1^* = (\sigma^{-2} + n\tau^{-2})^{-1}, \quad V_2^* = (\sigma^2 + n\tau^{-2})^{-1}$$

and where

$$\frac{\alpha^*}{1-\alpha^*} = \beta\sigma\left(\frac{\sigma^{-2}+n\tau^{-2}}{\sigma^2+n\tau^{-2}}\right)^{\frac{1}{2}} \exp\{-\tfrac{1}{2}(\mu-\bar{x})^2\,[(\sigma^2+n^{-1}\tau^2)^{-1} - (\sigma^{-2}+n^{-1}\tau^2)^{-1}]\} \qquad (5.1.4)$$

The density $g(\theta \mid \underset{\sim}{x})$ will be bimodal provided that $(\mu-x)^2$ is considerably greater than zero. Suppose we now wish to estimate θ using the conjugate loss function mentioned in Section 1. It is then easily checked that the corresponding expected loss function $E(\delta)$ satisfies

$$1 - E(\delta) = \hat{\alpha} f(\delta \mid \mu_1^*, V_1^* + k) + (1-\hat{\alpha}) f(\delta \mid \mu_2^*, V_2^* + k)$$

where $$\hat{\alpha} = \frac{\alpha^*(1+V_1^*k^{-1})^{-1/2}}{\alpha^*(1+V_1^*k^{-1})^{-1/2}+(1-\alpha^*)(1+V_2^*k^{-1})^{-1/2}}$$

and where $f(\delta)$ is defined above. We shall now consider the behaviour of the stationary values of $E(\delta)$.

Firstly suppose that either k or n is very large so that

$$\frac{V_1^*+k}{V_2^*+k} \simeq 1.$$

Then the conditions of Theorem 5.1 hold apart from the fact that the variances are only approximately equal. It can be shown that we can conclude that we have a cusp point approximately at the coordinates

$$(\alpha^*, (\bar{x}-\mu)^2) \simeq (\tfrac{1}{2}, 4(V_1^*+k))$$

If $(\bar{x}-\mu)^2 > 4(V_1^*+k)$ then the posterior expected loss bifurcates. Using the approximate symmetry of $E(\delta)$, the lowest minimum will be the one nearest μ if $\alpha^* > 1/2$ and the one nearest $\bar{x}$ if $\alpha^* < 1/2$. Thus if we use equation (5.1.4) we see that we choose the minimum closest to μ if and only if $(\mu - x)^2 < R$ where R is some complicated function of σ^2, τ^2/n and β. The trick now is to identify the minimum of $E(\delta)$ closest to μ as the "accept the null hypothesis" decision and the minimum closest to $\bar{x}$ as the "reject the null hypothesis" decision. In this way we can produce a Bayesian estimation analogy to the classical hypothesis test. It should be noted that this scheme suggests the acceptance of the null hypothesis when the Bayes decision δ is "near" and not equal to μ. However we feel that such a scheme is more in the spirit of hypothesis testing. In practice the classical statistician never expects his null hypothesis to be exactly right anyway.

If k and n are both small then we can no longer use the theorem. However it can easily be shown that we retain the cusp geometry for fixed k and in fact the bifurcation is more marked in the sense that two minima occur more quickly as we increase $(x - \mu)^2$. Notice that a difference between this type of estimation and the usual hypothesis testing estimate (or "testimate") is that in the former case the "critical region" defined by equation (5.1.4) depends on n and is not fixed size.

Example 5.2 *The Drunken Driver.* A study of the effect of drunkenness on driving skills was performed by Drew, Colquhoun and Long (1959). A graph of the data they obtained is given in Fig. 5.3. Later, an explanatory model of this data was put forward by Zeeman (1977).

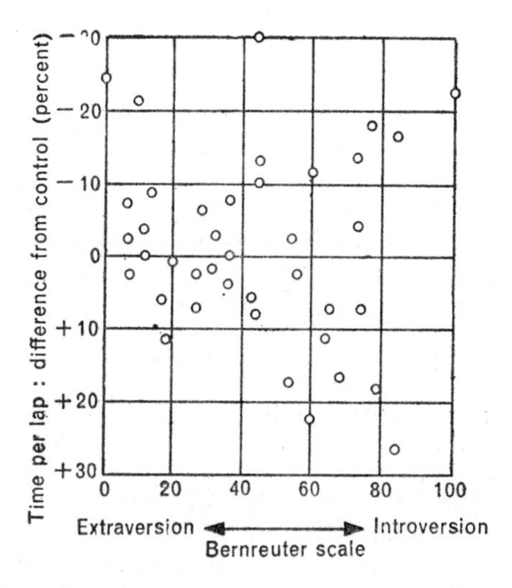

Figure 5.3. Mean time per lap (difference from control) for Dose 4 plotted against extraversion measured by the Bernreuter Scale.

Zeeman argued that a driver takes cues from his surroundings to estimate the speed of his car. When the driver drinks his integrative capacity goes down and he tends to pay more attention to either the faster speed cues or the slower speed cues. Hence instead of obtaining a unimodal belief distribution of his speed cues he obtains bimodal distribution. We can conveniently think of such a distribution of speed cues as a mixture of an overestimation distribution centred at $S + B$ and an underestimation distribution centred at $S - B$ where S represents his actual speed and $B > 0$. At his normal speed S_N we assume that the two distributions are weighted approximately the same. If he increases his speed $S > S_N$ then the proportion of overestimation cues to underestimation cues increases; similarly when he decreases his speed this proportion decreases. Our intention now is not to discuss the validity of this model but rather to show how we can use Bayes decision theory to link the model to the data given in Fig. 5.3.

For the sake of simplicity again, we will assume that the driver's distribution of speed cues when intoxicated can be expressed adequately as a mixture of two Normal distributions with the same variance and that his loss function is a conjugate one given in Section 1. Theorem 5.1 shows that this assumption is certainly not necessary but it makes the explanation of the model easier. We shall in addition assume that the introvert's loss function has parameter k which is very small so that he is very sensitive about the estimate of his speed being even slightly wrong and that the extrovert has a very large parameter k so that he tends only to be very concerned about gross errors in estimation. Thus k will be an increasing function of extroversion. According to Theorem 5.1 it will follow that the typical extrovert's expected loss function will have only one local minimum and this decision will be close to S. However, the typical introvert's expected loss function will have two local minima $d_1(S)$ and $d_2(S)$ (say) where

$$d_1(S) < S < d_2(S)$$

Suppose now that a particular introvert tends to pick out slightly more low speed cues than high speed cues when travelling ot his normal speed, S_N. For this introvert S_N will be near the local maximum of his expected loss function. The global minimum of his expected loss function will occur at $d_1(S_N)$ and so his estimated speed

(=his Bayes decision) $= d_1(S_N)$. Therefore he will deduce that he is driving too slowly because this estimate is less than his normal speed, S_N. Consequently he will accelerate. As his speed S increases smoothly, both $d_1(S)$ and $d_2(S)$ will increase smoothly, and so his estimated speed $d_1(S)$ will increase smoothly. Let S_1 denote the speed such that $d_1(S_1) = S_N$. However, he may never reach S_1 for the following reason. As he increases his speed, the proportion of high-speed clues will increase, causing the local minimum at $d_2(S)$ to decrease relative to $d_1(S)$. Let S_2 denote the speed at which these two reach the same level. If he increases his speed S past S_2, then his estimated speed wlll suddenly switch from $d_1(S)$ to $d_2(S)$. Hence there are two possibilities, according as to whether $S_1 \lesseqgtr S_2$.

Firstly, if $S_1 < S_2$, then when he reaches S_1 he will maintain a steady speed at S_1, going much faster than normal since $S_1 > d_1(S_1) = S_N$, while his estimated speed remains steadily at $d_1(S_1) = S_N$, so that he *thinks* he is driving at normal speed.

Conversely, if $S_2 < S_1$, then as soon as he reaches S_2 his estimated speed will suddenly jump to $d_2(S_2)$, and he will realize he is going too fast, because

$$d_2(S_2) > S_2 > S_N.$$

Therefore he will decelerate, but as soon as his speed drops below S_2 again, then his estimated speed will suddenly jump back to $d_1(S_2)$ again, and he will think that he is going too slowly because

$$d_1(S_2) < d_1(S_1) = S_N.$$

Therefore he will accelerate again and repeat the cycle. Thus his speed will oscillate around S_2, going much faster than normal since $S_2 > S_N$, while his estimated speed will continually jump to and fro either side of S_N.

On the other hand, if we choose a different introvert who this time picks out slightly more high speed cues than low speed cues when travelling at the true speed then by exactly analogous arguments so this individual's average speed will be far *less* than it should be. It follows that the slightest difference in the faculties between these introverts gives rise to widely divergent behaviour. It should also be noted that the argument implies that the more introverted a person is, the more his speed when intoxicated will differ from his normal speed. This is because for the same weighing parameter α, the smaller k the further the two local minima of $E(\delta)$ are apart. Of course whether this difference is too large or too small will depend on the ratio of high to low speed cues he identifies whilst travelling at the correct speed.

This example shows that by using Zeeman's model and Bayesian decision theory we would expect the sort of data discovered by Drew et al and given in Fig 5.3. Zeeman (1977) successfully fits a symmetrical cusp to the data when the time per lap scale is converted to a speed scale and thus vindicates our model.

Example 5.3 *The Behaviour of an Elected Decision-Maker.* An elected decision-maker M has two bodies of people to please. For the sake of argument the first group comprising n_1 of the electorate we shall call the radicals and the other group comprising n_2 of the electorate we shall call the conservatives. M's strategy is to maximize his expected support over the whole of the electorate by the decision δ that he will make. The success of this decision wlll depend on the level ψ of a particular process whose true value everyone will know only after M's decision is made. We shall assume, however, that M's information about this level at the time he makes his decision can be expressed in terms of a probability density function $f(\psi)$ on the real line which is symmetric about some value x and unimodal.

If ψ were known to take the value of $\psi \in \mathcal{R}$, then the radicals will accept M's decision δ if δ lies in the interval $(\psi + m_1 - b, \psi + m_1 + b)$. Conversely the conservatives will accept decision δ only if it lies in the interval $(\psi + m_2 - b, \psi + m_2 + b)$. We shall now investigate how M's best decision varies with m_1, m_2, n_1, n_2 and $f(\psi)$ when δ is chosen to maximize his expected support amongst the $n_1 + n_2$ electorate.

Firstly note that the expected support $S(\delta)$ for M when he makes decision δ is given by the equation

$$S(\delta) = n_1 + n_2 - n_1 E(\delta - m_1 - x) - n_2 E(\delta - m_2 - x)$$

where $E(y) = 1 - \int_{y-b}^{y+b} f(\psi + x)\, d\psi.$

It is easily checked that $E(y)$ is symmetric in y about 0 and

$$\lim_{p \to \infty} E(y) = 1$$

Therefore by Corollary 5.1.1, provided $E(\delta)$ is of type T, $S(\delta)$ is a function with cusps located only along the coordinates

$$(\delta, n_1, n_2, m_1, m_2) = (m, n, n, m + \eta, m - \eta)$$

where $m = x + \frac{1}{2}(m_1 + m_2)$ and η is the unique zero of E'' in $(0, \infty)$. Here the normal factor is $n_1 - n_2$ and the splitting factor $m_1 - m_2$. Thus we can make the following qualitative assessments of the situation. Let $\hat{m} = \frac{1}{2}(m_1 - m_2)$.

(a) If $\hat{m} < \eta$ then the requirements of the conservatives and the radicals will be similar enough for M to choose a point of compromise between the two groups.

(b) However if $\hat{m} > \eta$ then M will side with one of the two groups. If $n_1 > n_2$ he will choose δ^* close to $x + m_1$.

(c) It is possible for M to behave in a catastrophic way. For example suppose that $n_1 = n_2 + 1$ and that $\hat{m}$ is much greater than η. M will then choose a decision close to $x + m_1$. However, if the situation were to change slightly and two of the radicals changed allegiance to the conservative camp, then M's best decision would suddenly switch to a decision somewhere near the point $x + m_2$.

(d) Provided the regularity conditions of Theorem 5.1 are met, typically as $f(\psi)$ becomes more informative so η becomes smaller and as $f(\psi)$ becomes more diffuse so η will increase. This means that if M has a great amount of information about the level ψ he will tend to side with either the radicals or conservatives whereas if he is ignorant he will tend to compromise between the two groups.

Zeeman (1977) analyses similar types of models to the one above postulating the evolution of bimodal probability density functions of the support given to certain actions by a governor or governing body. However the link between the governor's "best" action and these densities is very tenuous and has been criticized. By looking at the expected support function instead this difficulty is co. pletely overcome.

Conclusion

This paper has indicated a wide range of models which exhibit discontinuities and has sketched out how it is possible to analyse such processes. With an underlying understanding of the mechanism generating sharp changes in behaviour it is possible to build predictive models which are robust to such changes. Quantitative models can then be built using the understanding. We feel that in the future this will increase the satisticians powers of prediction when dealing with such problems.

References

Chidley, J. (1976) "Catastrophe theory in consumer attitude studies", J. Mkt. R. Soc. 18.2, pp. 64–92.

Drew, G.C., Colquhoun, W.P. and Long, H.B. (1959) "Effects of small doses of alcohol on a skill resembling driving", Medical Research Council Memo Vol. 38.

Harrisson, P.J. and Stevens, C.F. (1976) "Bayesian Forecasting (with discussion)", J.R.S.S.B. Vol. 38, No. 3, pp. 205-247.

Ibragimov, I.A. (1956) "On the composition of unimodal distributions", Theory of Probability and Its Applications Vol. 1.

Kadane, J.B. and Chuang, D.T. (1978) "Stable decision problems", Annals of Statistics, Vol. 6, pp. 1095–1111.

Lindley, D.V. (1976) "A class of utility functions", Annals of Statistics, Vol. 4, pp. 1–10.

Smith, J.Q. (1978) Problems in Bayesian statistics related to discontinuous phenomena, catastrophe theory and forecasting", Ph. D. Thesis, University of Warwick.

Smith, J.Q. (1979) "Mixture catastrophes and Bayes decision theory", Math. Proc. Camb. Phil. Soc. Vol. 86, pp. 91–101.

Thom, R. (1972) "Stabilitè structurelle et Morphogènèse", Benjamin, New York; English translation by D.H. Fowler 1975.

Zeeman, E.C. (1977) Catastrophe Theory—Selected Papers 1972–1977, Addison-Wesley, Reading, Mass.

FIVE

Applicable Catastrophe Theory IV anorexia nervosa and its cure by trance therapy

E.C. ZEEMAN

Anorexia is a nervous disorder suffered mainly by adolescent girls and young women, in whom dieting has degenerated into obsessive fasting. It generally begins between the ages of 11 and 17, although it can start as early as 9 or as late as 30. It can lead to severe malnutrition, withdrawal and even death.

The proposed model is the joint work of the author and J. Hevesi, who is a psychotherapist specializing in anorexia. Hevesi has spent some 5000 hours during the last 5 years talking to over 150 anorexics and the model is based on his close observations. Of these 150 over 60 agreed to undertake his course of treatment, and of those treated he has achieved an 80% success rate of complete cure. His innovation is the use of trance-therapy. The Anorexic Aid Society in Britain recently conducted a survey of over 1000 anorexics, and the secretary of the society, Mrs. P. Hartley, who is a psychologist, writes: "I first read of Mr. Hevesi in several letters from patients who responded to my appeal for information about anorexia nervosa, and their experience re. treatment. These patients are the only ones who claim that they have recovered completely—i.e. those whose *attitude* to life has changed since undergoing Hevesi's treatment. They are not just eating properly (only the awful surface problem anyway) but living a full life as a *complete* personality." (her emphasis).

The advantage of using mathematical language for a model is that it is psychologically neutral; it permits a coherent synthesis of a large number of observations that would otherwise appear disconnected, and in particular enables us to place the trance states in relation to other behavioural modes. As yet the model is only qualitative, in the sense that the predictions that have been verified by observation have been qualitative rather than quantitative. Nevertheless it does provide a conceptual framework within which the theory could also be tested quantitatively by monitoring patients. Meanwhile we hope that it may not only give a better understanding of anorexia and its cure, but also provide a prototype for understanding of other types of behavioural disorder.

A striking feature of anorexia is that it sometimes develops a second phase after about two years, in which the victim finds herself alternately fasting and secretly gorging; the medical name for this is bulimia, and anorexics often call it stuffing or bingeing. If we regard the normal person's rhythm of eating and satiety as a continuous smooth cycle of unimodal behaviour, then we can interpret this second phase of bimodal behaviour, as a catastrophic jumping between two abnormal extremes. Therefore by the main theorem we can model the anorexic's behaviour by a cusp-catastrophe, in which she is trapped in a hysteresis cycle, as in Fig. 1.

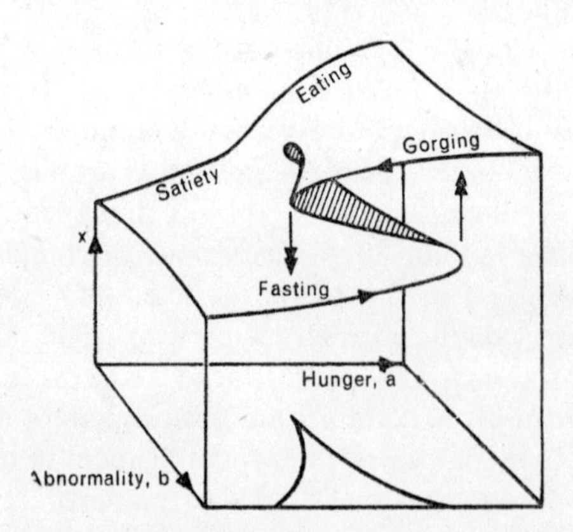

Figure 1 Initial behaviour model for anorexia.

Before we begin to analyse the model, we can immediately draw one important conclusion; *the victim will be denied access to the normal modes in between.* The denial of access to normal modes occurs already during the first phase of only fasting. Thus the main thrust of our approach will be to explain anorexia not as the complicated behaviour of a perverse neurotic, but as the logical outcome of a simple bifurcation in the underlying brain dynamics. If this is the case then catastrophe theory at once indicates a theoretical cure: if we can induce a further bifurcation according to the butterfly catastrophe, then this should open-up a new pathway back to normality. The practical problem is how to devise a therapy that will induce such a bifurcation, and this is what Hevesi's treatment achieves.

In Fig. 1 we have chosen hunger and abnormality as the two control factors (a, b). Hunger is the normal factor because hunger normally governs the rhythmic cycle between eating and satiety; there are various known methods for measuring hunger, but we do not yet know which will be best to use for quantitative testing. We postpone the discussion on the measurement of abnormality until later. Meanwhile to measure the behaviour, it would be necessary to find some psychological index that correlates with the scale of wakeful states shown in Fig. 2.

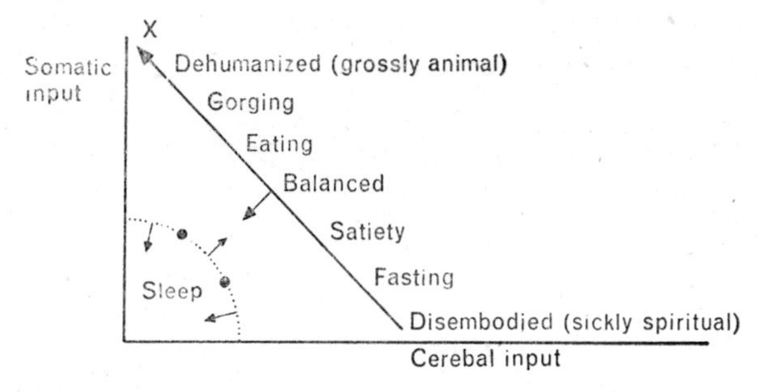

Figure 2. The x-axis measures both wakeful behaviour, and the relative weight given to the cerebal and somatic inputs to the limbic brain. The dotted line shows the boundary of the sleep basin, and the arrows show the movement of this boundary due to anorexia, leaving fixed the two nodal points.

What actually governs the behaviour is the underlying brain state, and if it is true, as MacLean suggests, that emotion and mood are generated in the limbic brain, then it is likely that x is measuring some property of limic states. Since the limbic brain receives both cerebal inputs from the neo-cortex, and somatic inputs from the body we might conjecture that x is some measure of the relative weight given to those inputs as shown in Fig. 2. Of course such a conjecture must remain speculative until it is confirmed or rejected by future brain research. Nevertheless the conjecture has already proved useful in explaining many of the symptoms of anorexia, and, what is perhaps more important, enabled us to identify what may be key operative suggestions in the therapy, as we shall see. Meanwhile the conjecture implies that the main neurological feature of anorexia is that during wakefulness the limbic brain is dominated either by cerebal inputs or by somatic inputs, while the balanced states have become unstable, and therefore inaccessible.

Before leaving Fig. 2 notice that it is 2-dimensional. From the psychological point of view the natural axes to use are x and y, which are inclined at 45° to the neurological axes, cerebal and somatic. Here x measures the different wakeful states, while y measures the difference between wakefulness and sleep, and y exhibits the familar healthy catastrophes of falling asleep and waking up. For a more complete model we ought really to use both the behaviour variables x and y, 5 controls, and a 7-dimensional catastrophe called E_6. However this is beyond the scope of this paper, and so for simplicity of presentation we shall sacrifice y and use only x.

We now introduce a third control factor c, which will play the role of the bias factor in the butterfly catastrophe. Define c to be *loss of self-control*, measured by loss of weight. Geometrically the effect of bias is to swing the cusp to and fro. The resulting effect on the behaviour surface is shown in Fig. 3.

During the first phase of the disorder the anorexic is firmly in control of herself, and so $c < 0$ as in Fig. 3 (i). The normal person has learnt to perform the regular smooth cycle at the back, socially structured by mealtimes. The anorexic however finds herself trapped on the lower sheet at the front by the abnormality; in other words the limbic brain oscillates continuously in states underlying a fasting frame of mind all the time she is awake, even when she goes through the motions of eating. The frame of mind is predominantly cerebal, and the victims often speak in terms of "purity"; it tends to smother instincts and produce excessive verbalization. During the first phase victims often deny being ill, and refuse treatment.

Then as the anorexic gradually loses weight, she gradually loses control of herself; the bias factor c gradually increases, causing the cusp to swing gradually to the left, as in Fig. 3 (ii). How far the cusp will eventually swing in relation to the cycle will depend upon the individual. If it swings sufficiently far for the right-hand side of

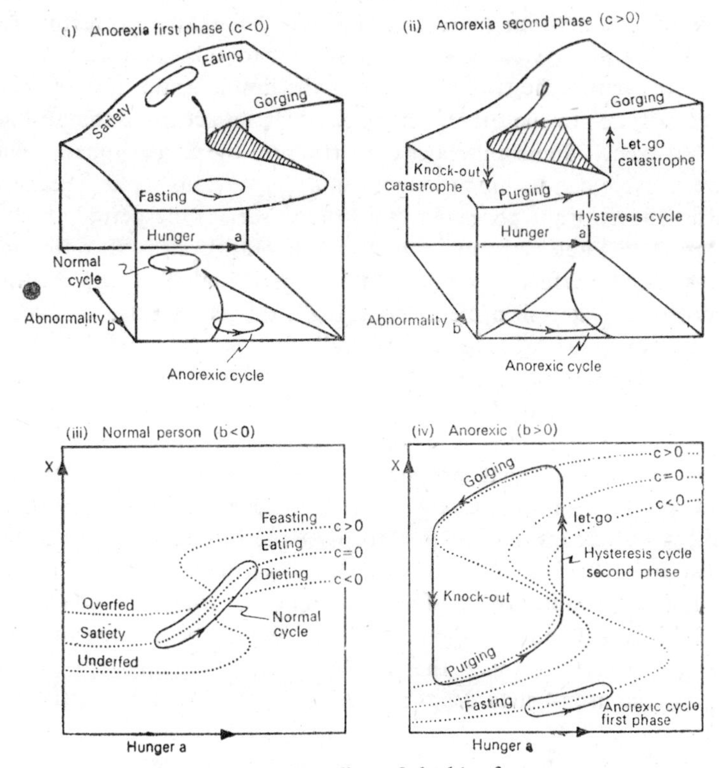

Figure 3. The effect of the bias factor, c.

(i) Anorexia first phase ($c < 0$). Strong self-control swings the cusp to the right, and abnormality displaces the normal cycle into fasting.
(ii) Anorexia second phase ($c > 0$). Loss of self-control swings the cusp to the left, causing the anorexic to jump into the catastrophic hysteresis cycle of alternately gorging and purging.
(iii) Normal person ($b < 0$). Changes in bias modify the behaviour slightly.
(iv) Anorexic ($b > 0$). Changes in bias modify the behaviour dramatically.

the cusp to cross the right-hand end of the cycle, then this will cause the sudden onset of the second phase. For now, instead of being trapped in the smooth fasting cycle on the lower sheet, the victim finds herself trapped in the hysteresis cycle, jumping between the upper and lower sheets. The catastrophic jump from fasting to gorging occurs when she "lets go": in the victim's own language, she watches helplessly as the apparent "monster inside herself" takes over, and devours food for several hours. Some victims vomit and gorge again, repeatedly. The catpstrophic jump back occurs when exhaustion, disgust and humiliation sweep over her, and she returns to fasting for a day or several days. Some anorexics refer to this as the "knock-out". At each of the two catastrophes the limbic brain jumps from one set of states to the other, denying the victim access to the normal states between. Some anorexics even ritualize the catastrophes. The hysteresis cycle can be much longer than the previous cycle, because the after-effects of the gorge tend to prolong the fasting period.

Figs. 3 (iii) and (iv) show how the different cycles fit onto the sections of the surface. Notice that we have labelled the fasting period of the hysteresis cycle as "purging"; this is because it occurs at a different value of x to the "pure" fasting of the first phase. Indeed the two limbic states underlie quite different frames of mind; fasting is cereballly dominated, not allowing food to enter, while purging has the somatic element of getting rid of bodily contagion.

It is not known what proportion of anorexics switch into the second phase. Sometimes the switch occurs after a hospital treatment with drugs that are used to persuade the starving first phase anorexic to eat. If the effect of such drugs is to reduce cerebal inputs to the limbic brain in favour of somatic inputs, which is consistent with observed side-effects, then the drugs would be reinforcing the bias and therefore *causing* the switch. Thus the long-term harm caused by the use of such drugs may be greater than the short-term benefits.

Now we come to the cure. The strategic problem is how to persuade the anorexic to relinquish her abnormal attitudes, but this cannot be done directly. Therefore the problem is how to break the vicious circle:

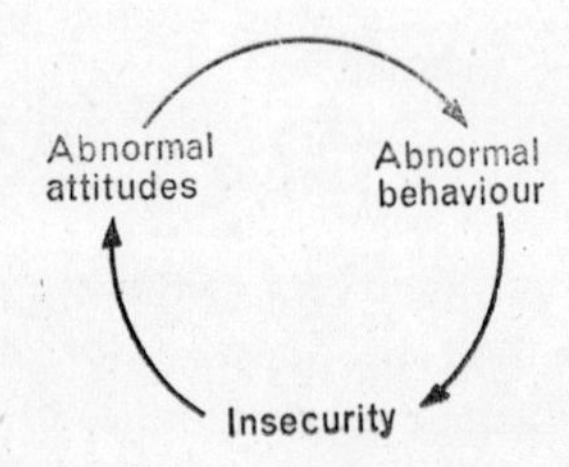

The idea is to break in at the behavioural corner, by creating a third abnormal behaviour mode, during which the insecurity can be treated with reassurance. We will later show how this in turn causes a catastrophic collapse of the abnormal attitudes.

The new behaviour mode must lie between the abnormal extremes if it is going to provide a context within which reassurance can be effective. Therefore the butterfly catastrophe in Fig. 4 tells us the geometric relationship that this mode must have (i.e. the dynamic relationship that the underlying brain states must have) in relationship to the existing modes.

Meanwhile Fig. 2 shows that we must look for such a mode in the twilight zone between waking and sleeping for the following reasons. In the huge dynamical system modelling the limbic states of a healthy person, sleep is an attractor (i.e. a stable oscillation) with a stable boundary to its basin of attraction, separating it from wakefulness. In Fig. 2 we have symbolically indicated the boundary by a dotted line. In the anorexic the boundary becomes fuzzy because the basin is being shifted, as indicated by the arrows periferally the basin is being eroded by the increasing stability of the abnormal extremes, while in between it is being enlarged by the decreasing stability of the balanced states. These changes cause the sleeping patterns to be disturbed: sleep is fragmented, shifted around and edges of the fragments become fuzzy; the anorexic goes to bed late, wakes at night sleeps little, finds herself lounging about in her night clothes. Moreover, for exactly the same mathematical reason that temporary lakes sometimes appear on the boundaries of river basins near the nodal points in between erosion and growth, so fragile attractors may appear at the boundary of the sleep basin, particularly near the two nodal points marked in Fig. 2. Therefore the anorexic finds herself spontaneously falling into fragile trance-like states, in the twilight zone between waking and sleeping, between dreaming and perceiving. At the somatic node these trance-like states are filled with thoughts about food, and lists of food, while at the cerebal node they are shot through with schemes and plans how to get through the day, how to manage social occasions and avoid set mealtimes, their preparation and aftermaths, shopping, cooking and washing up.

It is these confused trance-like states that are utilized by the therapist; *therapy builds upon naturally occurring processes.* Hevesi's treatment consists of about 20 sessions of trance-therapy over a period of 6 to 8 weeks, each lasting 2 to 3 hours. When the sufferer asks for help, the therapist begins by pushing aside the inconclusive and confusing contents of these states, pushing them away in their respective directions so as to create a new more balanced trance.

Because of the state of the sufferer quite casual remarks can carry the force of suggestions, and thus the operative suggestions are actually made quite marginally, almost incidentally. Firstly a casual but firm announcement is made at the beginning (and abhered to throughout the treatment) such as "I don't care what you eat—we are not going to talk about eating or food", because this reduces the somatic input. Secondly after the formal step of going into the trance, a suggestion is made such as "Let your mind drift—don't think—look" because this reduces the cerebal input.

Thus the patient's mind is cleared of both food and scheming, and is free to look at itself. By contrast when she is fasting she is looking all the time at the outer world with anxiety, and when she is gorging she is overwhelmed by this same world, but during trance she is cut-off and isolated. By suspending the threats, the rules, the resistance and the hunger the trance gives temporary freedom from anxiety. She is able to look at the products of her own mind, and contemplate its image and memories. In this state she is open to reassurance, and, more importantly, *able to work out her own reassurance*.

The more the patient practises trance, the easier it becomes; rainforcement causes an increase in stability of the new attractor, and an enlargement of its basin of attraction. The trance states begin to emerge as the new middle sheet of Fig. 4 (i). Therefore we introduce the last control factor, d, as *reassurance*, measured by time under trance.

Summarzing the four control factors:

a : normal factor: hunger.
b : splitting factor: abnormality, (measurement discussed below).
c : bias factor: loss of self-control, measured by loss of weight.
d : butterfly factor: reassurance, measured by time under trance.

Going into trance is a catastrophic jump from the lower sheet (because therapy usually takes place during the fasting part of the cycle) onto the middle sheeet. Therefore the patient tends to fall into trance. What causes this jump? In fact the jump has two components, a relatively small one in the x-direction, and a larger one in the y-direction towards sleep, which is the second behavioural variable of Fig. 2 that we have omitted from Fig. 4 for simplicity.

And it is not caused by a reduction in the abnormality, b, but by an increase in drowsiness, which is the fifth control factor, again omitted for the same reason; this is the only point where the simplification has caused a slight geometrical inaccuracy in our pictures.

Coming out of the trance is another catastrophe, and causes the reverse jump back onto the lower or upper sheet, depending upon whether the left or right side of the pocket is crossed, as shown in Fig. 4 (ii) The patients confirm that when they awake from the first few trance sessions they find themselves sometimes in a fasting and sometimes in a gorging frame of mind.

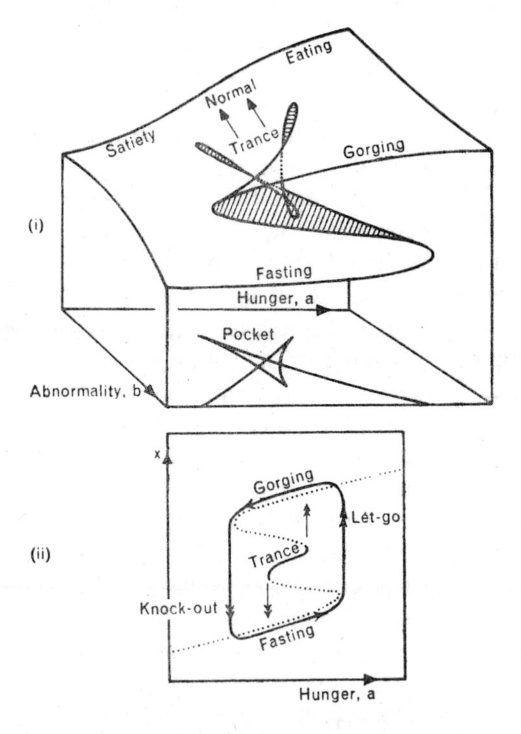

Figure 4. The effect of the butterfly factor, $d > 0$.

(i) When therapy starts the trance states appear as a new triangular sheet of stable behaviour over the pocket in between the upper and lower sheets. The new sheet opens up a pathway back to normality.

(ii) The trance states sit inside the hysteresis cycle; initially they are fragile, and coming out of trance is a catastrophic jump into either a fasting or a gorging frame of mind.

We now come to what happens during the trance. As the therapy progresses Hevesi's patients report that they experience three phenomena, of which the third is observable from the outside. Of course what the mind sees in trance it has put there, and interpreted in its own fasnion, even though the images will naturally be made according to past experience, and the feelings will be such as are stored up from the past. The experience in trance may be compared to the steps in which an actor approaches a role. The first step is to envisage the part in a few simple strokes or characteristics; the second is to hear the lines the character is allotted to speak (say in a first reading through of the script); the third is to get into the part and play it, to act in front of an audience.

The first phenomenon is an experience of herself as a double personality; one personality is usually described as the "real self" and the other is called various names by different patients such as "the little one, the imp, the demon, the powers, the spirit, the voice" or merely "it". Possibly the suggestion by the therapist to look rather than to think may prepare the way for the appearance of "persons", but usually the latter appear by themselves, and we shall argue below that the patient is in fact giving a logical description of herself. It is the voice or however she describes it, who is apparently issuing the prohibitions over food: "The little one says I musn't eat". Typically the first appearance may occur about the third session: "I've got a voice", and then perhaps a couple of sessions later "This is the first time the voice has spoken in public".

The second phenomenon is an apparent transfer of important messages between the two personalities, such as the real self promising to "pay attention" to the little one, reassuring the little one that she "will not be forgotten", while the latter in return agrees to relax the prohibitions. Sometimes the little one is symbolically given a gift, such as a teddy-bear that she once longed for and never got.

The third phenomenon is a "reconciliation" or "union" or "fusion" of the two personalities, a "welcome possession" as opposed to the earlier malignant possession. Typically "She is coming out", or "She is very near the front", and then "the voice just seems to be part of myself". This third phenomenon is accompanied by a manifestation that can be witnessed by the therapist, such as speaking with a strange voice, and usually happens after about two weeks in around the seventh session (depending of course upon the individual).

When the patient awakens from this particular trance, she discovers that she has regained access to normal states, and is able to eat again without fear of gorging: she speaks of this moment as a "rebirth". Therefore during this trance the cure has taken place, a catastrophic drop in the abnormality, b, which we shall explain in a moment. Thus the trance states have opened up a path in the dynamics of the brain back to normality, indicated by the arrows in Fig. 4 (i). Subsequent trance sessions re-enact the reconciliation in order to reinforce normal states and buffer them against the stresses of everyday life. At the same time the trance technique is itself reinforced, so as to provide a reliable method of self-cure, should the patient ever need to use it again at a later date.

Having dealt with the behavioural point of view, we now turn to the heart of the problem: What causes anorexia? Why can most slimmers diet without becoming anorexic? Why is there such a slow insidious apparently irreversible escalation of the disorder? How can we measure the abnormality, b? Why is the resulting neurosis/psychosis so rigid? How can there possibly be such a dramatically sudden cure?

We shall add one more cusp catastrophe to the model that will answer all these questions except the first. The reason that it cannot answer the first is that the model refers to what can be observed, whereas the original causes are probably hidden much earlier in childhood. We can offer an analogy, which may give some insight, but is not strictly part of the model. The metaphor is to describe anorexia as an "allergy to food"; of course it is *not* an allergy, but it does show some qualities similar to those of an immune system being set-up, switched-on, and inducing exposure-sensitivity. The origin of anorexia may occur in early childhood, when, perhaps for want of love or due to the inability to obtain the attention that it needs, the child retires into its shell; in other words the personality sets up an immunity against disappointment by turning inwards, and leaving the shell to act out the game of life. This immunity works well enough until the shell begins to grow and get out of hand, when the encapsulated core of the personality finds that it can no longer manage. It is then that the anorexia is switched-on, instinctively identifying food as the cause of growth. This may be why anorexia so often begins at the onset of puberty, or after a period of obesity. From now on the victim is exposure-sensitive to food, and being presented with food raises deep-seated anxieties. The logical reaction is to avoid stressful situations, and so the core begins to issue prohibitions to the shell concerning food. Consequently the victim begins to feel an urge to avoid food, which she cannot explain; when she attempts to explain it she tries to capture in words some quality of the urge: e.g. "the little one" is a recognition of its origins in childhood, "the imp" describes its bad quality, "the voice" its unidentifiableness. Usually such attempts are met with disbelief, and she soon stops trying to explain.

Our metaphor breaks down when the anorexia begins to escalate. This can be observed, and so can be put into the model, as follows. Increasing insecurity is observed, associated with a gradual escalation of abnormalities over food. A typical escalation might include the following stages, but of course each anorexic will differ in the details of her own particular escalation.

	tummy-aches at school
escalating	give up carbohydrates
stages of	give up cooked meat
abnormality	elaboration of diet (weight watching, calorie counting)
over	excessive diet (e.g. only cheese, peanuts, black tea)
food	excessive activity
	deception (e.g. secret purging after each meal)
	manipulation (of rest of family)
	nocturnal eating only
b ↓	"I don't need to eat"

The important observations for our purpose are (1) there is an escalation of stages; (ii) at each stage a bimodal attitude is possible, normal or abnormal; (iii) when the anorexic reaches each stage, she will already have adopted normal attitudes towards all the previous stages, but as yet maintains normal attitudes towards the subsequent stages; (iv) increasing insecurity is associated with the escalation. Interpreting these facts geometrically gives the graph in Fig. 5 (i) showing the normality of attitude as a function of insecurity level, i, and abnormality stage, b. Abnormal attitudes begin at stage b_0, when insecurity has reached level, i_0. By the time insecurity has reached level i_1 the anorexic will have adopted abnormal attitudes towards stages up to b_1, but will so far have maintained normal attitudes towards stages beyond b_1. Thus b_1 measures the level of abnormality. Then, as the insecurity increases, so does the abnormality, following the curve in the horizontal plane, confirming that the onset of anorexia is a continual escalation by a succession of little catastrophes, little changes of attitude. Moreover as the disorder deepens, the individual catastrophes become bigger, and the attitudes towards earlier stages more abnormal (carbohydrates are at first avoided, and later feared).

We now appeal to the main theorem. The stability of memory

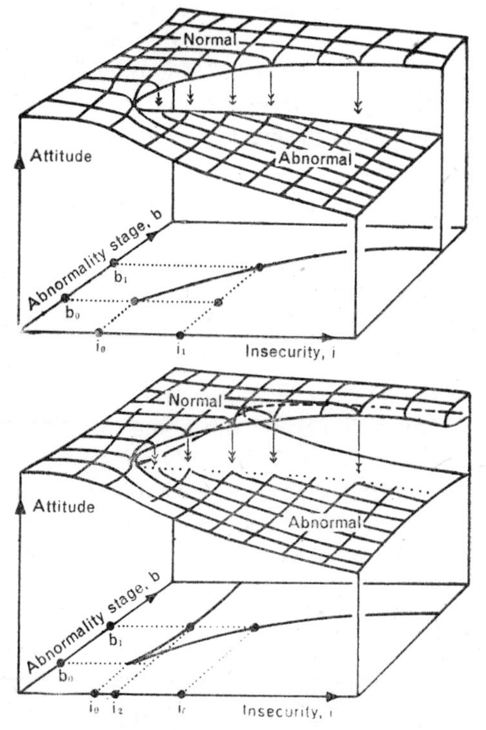

Figure 5. Abnormal anorexic attitudes

(i) The graph showing the escalation of anorexia by a succession of little individual catastrophes, as abnormal attitudes are adopted towards each stage.

(ii) The graph embedded in a cusp-catastrophe. The left branch of the cusp indicates how far the insecurity must be reduced in order to effect a cure.

and habit implies the existence of an implicit dynamic that holds the attitudes stably on the graph. The existence of a dynamic allows us to deduce that the graph is part of a cusp catastrophe. The right branch of the cusp marks the points where the stability of normal attitudes breaks down, and the attitude switches to abnormal, causing the gradual escalation of the disorder, while the left branch marks the points where the stability of abnormal attitudes breaks down, and the attitude switches back to normal. Thus the left branch (whose existence is a consequence of the theorem) predicts the possibility of complete cure.

The left branch also explains why the disorder is rigid and seemingly irreversible, as follows. Suppose that after reaching i_1 the insecurity level drops again. Then the abnormality will *not* drop, but will stay fixed at b_1 until the insecurity has dropped to i_2 (where i_2 is given by the intersection of the line $b = b_1$ with the left branch), because all the abnormal attitudes will be held stabily on the lower sheet, up to the boundary of that sheet. Then, as the insecurity drops from i_2 to i_0, there is a sudden rush of attitudes switching back to normal, along the left brench of the cusp.

Notice that the worse the anorexia is, the more rigid and irreversible it is, because as b_1 increases, so does the length of the interval $i_1 - i_2$, and hence the greater the reduction in insecurity that must be achieved before any improvement can take place. The model also explains why rasoning with the victim about her eating habits may be worse than useless, because it can only reinforce her insecurity and cannot change her attitudes; what is needed is the more fundamental reassurance about the source of insecurity. But the anorexic is not open to such reassurance while she is obsessed with, and transparently aware of her abnormal behaviour. Hence the utilization of the trance states, in order to give a temporary freedom from that obsession and awareness. Fig. 6, which is deduced from Fig. 5, shows how the therapist under these conditions can, by gently reducing the insecurity, trigger a dramatically swift catastrophic cure. Fig. 6 also illustrates the difference between slimming and anorexia, showing how a quantitative difference in the initial insecurity can lead to a qualitative difference in the eventual outcome that will enable the slimmer to achieve her slimness without danger, but prevents the anorexic from escaping from her prison without help.

We have used the word "cure" in the sense of the fundamental change of attitude to life, referred to in the very revealing testimony of Mrs. Hartley quoted at the beginning: the actual physical recovery from the accompanying malnutrition and amenorrhoea will then follow naturally over the next few months. It is doubtful if this type of

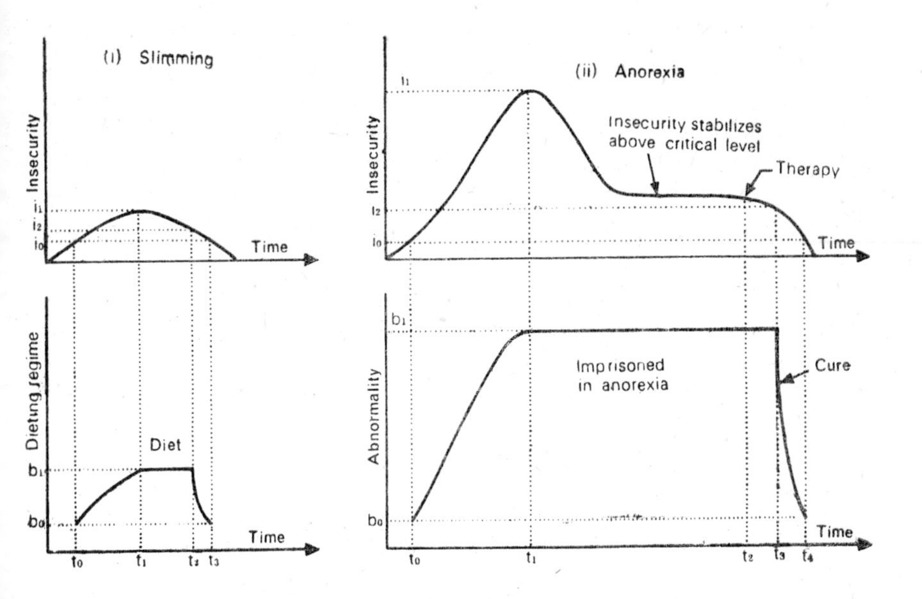

Figure 6. Comparison between slimming and anorexia.

(i) The *slimmer* anxious about her size, reaches insecurity threshold i_0 at time t_0, and therefore begins to diet with regime b_0; reaches maximum insecurity i_1 at time t_1, and therefore stabilizes dieting regime at strictness level b_1; finds, as dieting succeeds, that insecurity drops to the critical level i_2 by time t_2, and therefore rapidly relaxes her dieting regime; finds that insecurity drops to threshold i_0 by t_3, and therefore gives up dieting.

(ii) The *anorexic* begins the same, except that due to deep-seated anxieties reaches a much higher maximum insecurity i_1, causing her abnormality to develop and stabilize at level b_1; is prevented from reducing her insecurity to the critical level i_2 by the feedback from the abnormal behaviour forced on her by the anorexia, and therefore remains locked in the disorder; begins therapy at time t_2, which, by reassurance during trance, reduces the insecurity to the critical level i_2 by t_3, thereby effecting the catastrophic cure by t_4.

cure could be achieved while administering drugs that disrupt cerebal activity, because the recapturing of the whole delicate network of normal attitudes must depend not only upon reassurance, but also upon harnessing the full power of the cerebal faculties rather than suppressing them.

One of the most interesting points made by Paul MacLean is that the limbic brain is non-verbal, being phylogenetically equivalent to the brain of a lower mammal. Therefore the problem of describing its activity in ordinary language is like trying to describe the conversation of a horse; no wonder anorexics have difficulty in explaining their symptoms. The patient can perceive that certain subsets of states are connected, and have boundaries, and so for her the most logical approach is to identify those subsets as "dissociated subpersonalities", and give them names. In the model these subsets are represented by the different sheets, and the structural relation between them is defined by the unique geometry of the catastrophe surfaces. Therefore we may identify those sheets with the patient's descriptions of her subpersonalities. For example the upper sheet of Fig. 1 is often called the "monster within", and the lower sheet the "thin beautiful self". When she goes into trance the reduction of sensory input causes a shift in focus, from the close-up to the long-distance, from the immediacy of mood and behaviour to the long-term perspective of personality and insight. In terms of the model there is a shift from the perception of the states represented by the sheets of Fig. 1 to those of Fig. 5. Therefore the "monster" and "thin self" recede in importance, and are replaced by the "real self" the "little one" corresponding to the normal and abnormal sheets of Fig. 5. More precisely it is the dynamic holding the attitudes stably on the abnormal sheet that is dimly perceived and interpreted as "prohibitions" by the voice or as "malignant possession" by the little one. The "reconciliation" refers to the left branch of the cusp, which marks the boundary of the abnormal sheet, when the stability breaks down and the catastrophic cure takes place. Thus the apparent nonsense spoken by some patients makes perfectly good sense within the framework of a complete model.

Finally comes the question of how the model can be tested scientifically. It already satisfies Thom's criterion for science, because by its coherent synthesis it reduces the arbitrariness of description. Furthermore it has survived a number of qualitative experiments between myself and Hevesi of the following nature. From the mathematics I would make some prediction, or get depressed about some failing of the model, and then when we next met Hevesi could confirm the prediction, or confirm that what I had thought to be a failing was in fact another correct prediction. Let me give some examples. The mathematics predicted the location of the trance state as the middle sheet of the butterfly. However at one stage I thought the model had failed because bias destroys the middle sheet, meaning that for those patients the trance was not accessible, but to my surprise Hevesi revealed that he found very confirmed fasters or very confirmed bingers more difficult to cure. Another prediction of the mathematics was the qualitative difference between the "fasting" and "purging" frames of mind, illustrated in Fig. 3 (iv); the correctness of this prediction concerning the operation of the bias factor gave further evidence in favour of using the butterfly catastrophe.

Perhaps our most striking experiment concerned the finding of the operative suggestions. Hevesi says that the trance is not like hypnosis, because the therapist does not attempt to control the patient. I was curious to know what he actually did during the trance, but when I asked him he maintained that he did not do very much. Meanwhile the mathematics was insisting that we ought to look at the underlying neurology as well as the psychology, even if only implicitly, in order to locate the dynamic; consequently we formulated the conjecture about inputs to the limbic brain in terms of MacLean's theories. It was only then, after watching himself with new eyes, that Hevesi was able to lay his finger on the operative suggestions that were reducing those inputs. Thus the model facilitated the communication of the therapeutic technique.

To test the model quantitatively would require monitoring a patient in different states; the prediction would be that the psychological data would give catastrophe surfaces diffeomorphic to those extracted from the accompanying neurological and physiological data, with the same bifurcation set. Different patients would have diffeomorphic bifurcation sets, or parts of them diffeomorphic.

SIX

Discussion, Questions on Catastrophe Theory as answered by

E.C. ZEEMAN*

Q 1 Why is the theory called Catastrophe Theory? Who first used the word catastrophe and why? Do you think it expresses the basic concepts correctly or do you think it can be misleading? Can you suggest some other words to express the concepts in a better way?

E.C.Z. The theory is called "Catastrophe Theory" because Réné Thom wanted to emphasize discontinuity (contrast: Bifurcation theory). The name was given by Thom. When he found to his surprise the finite classification, he decided to call the classes elementary catastrophes. Non-elementary catastrophes arise in non-gradient dynamics. They are not classified so far. Examples are Hopf Bifurcations or Limit Points. In Hamiltonian Dynamics we get an infinite sequence of subharmonics, and the limit is a generalized catastrophe. It does express the basic idea of the surprise that continuous causes can give rise to discontinuous effects. But it is misleading because not all mathematical catastrophes are disastrous. No, there are no other words.

Q 2 What are the limitations of the theory?

E.C.Z. The limitation is that the dynamics is gradient like. Secondly the singularities are only local. There are very few global theorems as yet.

*As recorded by Purabi Mukherji, Department of Mathematics, Jadavpur University, Calcutta.

Q 3 Is it necessary to supplement its results with quantitative models based possibly on observations and experiments?

E.C.Z. If one has to apply it in science, then one has to predict and test it.

Q 4 Please give some cases where it has led to (i) new insights and (ii) new guidelines for action. What do you consider are the two most outstanding applications of Catastrophe theory?

E.C.Z. There are plenty of papers (about 200) which use catastrophe theory. The most outstanding application has been in the field of Caustics, viz., the theory regarding the twinkling of stars.

Q 5 What is the present position of generalization of Thom's theorem when $k \geqslant 4$.

E.C.Z. The topological classification in n dimensions is not yet solved but the best look in the direction are the theories of topological stratification by C.T.C. Wall in Liverpool. Meanwhile the lists of elementary catastrophes in higher dimensions are the same as Arnold's lists of normal forms of Lagrangian singularities. [A'rnold has published many papers in Russian which are available in English translations.]

Q 6 What is the relation between Catastrophe and Bifurcation theories?

E.C.Z. Historically Bifurcation theory considered only one control variable, and emphasized on the analytic complexities of the non-manifold arising from an infinite dimensional state-space. On the other hand, Catastrophe theory considered several control variables and a finite dimensional state space and emphasized the geometrical simplicities of the manifolds arising from stability and classification. Given any non-manifold of equilibrium states the catastrophe theorist wishes to "*unfold it*" to a manifold; this gives an insight into all possible local perturbations. The correct attitude is of course to combine the power of both the theories. The mathematical achievements of catastrophe theory are to stimulate development of the theories of singularities, unfoldings, stratifications and the preparation theorem.

Q 7 Apart Thom's theorem, what are the other main theorems of catastrophe theory?

E.C.Z. One cannot tell where the catartrophe theory finishes and all the other large mathematical theories listed above begin. The best attitude is to regard catastrophe theory as a catalyst stimulating a large amount of pure mathematics researches across the disciplines of differential topology, differential equations, algebraic geometry, commutative algebra, functional analysls and global analysis. In the years to come, the name catastrophe theory will disappear, but the theorems in these branches of mathematics will proliferate and the application of geometrical techniques of modelling in the sciences will increase.

Q 8 What are the pure mathematical problems in catastrophe theory waiting to be solved?

E.C.Z. (i) The global theory of singularities. For example, if we know the singularities in the boundary, what do we say about the interior.

(ii) To develop the topological stratification of function spaces and jet-spaces.

(iii) To put Arnold's definition of *normal form* on a rigorous basis, and to prove density theorems for his normal forms in dimensions 6.

(iv) To develop a catastrophe theory under group action, e.g. Lorentz group. If we do the whole singularity theorem subject to the Lorentz group (or other symmetric groups), then we get completely new singularities.

(v) To develop a similar theory using contact equivalence instead of right equivalence. $f \sim f'$ if $f = f'_g$, $g = G$

germs diffoes R^n 3, 0 ↑]

This is suitable for analysing the zeros of functions (or maps) as opposed to sigularities of maps. It applies to non-gradient dynamics. For example, the equilibrium of a dynamic = the zeros of the map

$$M \xrightarrow{v} TM]$$

Q 9 It has been stated that catastrophe theory has large pretensions and small achievements. What is your reaction?

E.C.Z. It would be a useful mathematical tool in science. Most mathematicians working in catastrophe theory feel the same way. It is the popular press that has mistakenly claimed great achievements. And it has two critics in particular, viz.—Sussman and Smale, who have blamed mathematicians for the statements in the popular press. Most other critics quote Sussman and Smale. In my opinion the controversy is a temporary storm in a teacup.

Q 10 How can one investigate catastrophe situations in a dynamical system represented by autonomous differential equations of nonlinear type in n state variables and m parameters? Will you please indicate the approaches to study such situations where they exist ?

E.C.Z. The non-gradient dynamical systems are as yet unsolved. Great advances in understanding have been made during the last fifteen years, notably by the American school [Smale and his students] and the Russian school [Arnold, etc.].

Q 11 Does hysteresis phenomenon indicate catastrophic situation?

E.C.Z. Yes. Very likely.

Q 12 If a nonlinear dynamical system possesses number of multiple-steady states (which usually occur in biochemical network analysis), can one predict the qualitative behaviour of such state variables possessing such states?

E.C.Z. Not without further information. If there is a parameter value that coaleses all these multiple states into one single state then the neighbourhood of the parameter may be describable as the unfolding of that single state.

SEVEN

Applications of Catastrophe Theory to phase transformations

C.K. MAJUMDAR

A phase transition is a cooperative phenomenon in which small continuous variation of interparticle forces and external conditions induces a sudden catastrophic change. All material substances show such transitions—there are certain universal features in them, and their mathematical description must involve similar mathematical singularities.

We shall briefly examine the connection of these with the work of R. Thom [1], E.C. Zeeman [2] and their collaborators purporting to provide a classification of possible mathematical singularities.

I. Generalized Free Energy Functional

A simple but sufficiently general description of continuous phase transitions starts with an introduction of an order parameter η. On one side of the transition is the ordered phase, characterized by non-vanishing values of η. Identification of the order parameter in a particular transition requires insight into its nature. For example, in the gas-liquid transition, the order parameter is density; in a ferromagnet, it is magnetization density.

The kinetics of phase transformations can be investigated with light scattering or resonance techniques. Conventionally, evolution equations for the order parameter η are written down for describing the kinetics [3]. A typical equation is the Landau Khalatnikov equation

$$-\partial \eta/\partial t = D\, H[\eta] \equiv D(\delta \phi[\eta]/\delta \eta) \qquad (1)$$

D is a phenomenological kinetic coefficient. $H[\eta]$ is an operator that drives the transition. When $\partial\eta/\partial t = 0$, the system is in a stationary state or in equilibrium. The equilibrium configurations are governed by maximum or minimum principles, e.g., entropy is maximum or free energy is minimum. This leads to the second equality of Eq. (1). Stationary values of the generalized free energy functional $\Phi[\eta]$ correspond to stationary states.

Physicists write down Φ from intutive or heuristic arguments or assumptions of simplicity. Thermodynamics does not tell us how to write such functionals, only stationary values are thermodynamically accessible. *It appears that the catastrophe theory might supply a guiding principle by which the functional can be written down in some cases.*

II. Thom's Classification Theorem

Consider a system described by a finite set of variables $x, y, z, \ldots$, and controlled by a second finite set of variables $a, b, c, \ldots$. The system is governed by an energy functional E dependent on $x, y, z, \ldots$ and $a, b, c, \ldots$. For fixed control parameters, the system takes up equilibrium values of $x, y, z, \ldots$ corresponding to stationary values of E. We ask: if we vary the controls $a, b, c, \ldots$ what kind of jump discontinuities can occur? Thom's theorem now states that with suitable choice of coordinates, the system can exhibit only seven types of singular behaviour when the system is described by at most two variables under at most four control parameters. These seven are called "elementary catastrophes" and their energy functions are given in Table 1.

Table 1. Energy function of elementary catastrophe (max. no. of state variable 2 : max. no. of control parameters 4).

Catastrophe	*Energy Function*
Fold	$\frac{1}{3}x^3 + ax$
Cusp	$\frac{1}{4}x^4 + \frac{1}{2}ax^2 + bx$
Swallow tail	$\frac{1}{5}x^5 + \frac{1}{3}ax^2 + \frac{1}{2}bx^2 + cx$
Butterfly	$\frac{1}{6}x^6 + \frac{1}{4}ax^4 + \frac{1}{3}bx^3 + \frac{1}{2}cx^2 + dx$
Huperbolic umbilic	$x^3 + y^3 + ax + by + cxa$
Elliptic umbilic	$x^3 - 3xy^2 + ax + by + c(x^2 + y^2)$
Parabolic umbilic	$x^2y + \frac{1}{4}y^4 + ax + by + cx^2 + dy^2$

III. Landau Theory of Phase Transition

We consider the simplest situation of Landau theory: ferromagnet with uniform magnetization in a uniform field H. The order parameter η is the magnetization density M.

Then the functional Φ is written as

$$\Phi(M) = \Phi_0 - MH + aM^2 + bM^4 \tag{2}$$

Thermodynamics is governed by

$$\frac{d\Phi}{dM} \equiv 2aM + 4bM^3 - H = 0 \tag{3}$$

For $H = 0$, this has the solutions

$$M = 0 \tag{4}$$

$$M = (-a/2b)^{1/2} \tag{5}$$

When $a > 0$, (5) leads to imaginary M and is unacceptable. The minimum of free energy is for $M = 0$. When $a < 0$, the minimum is attained for (5). At the critical temperature, a changes sign continuously, and the system passes from a state with vanishing magnetization to one with a non-vanishing magnetization. Equation (2) corresponds to the cusp catastrophe. The system is controlled by the external magnetic field H and the temperature T.

We must ascertain the behaviour of a and b as function of temperature. Following Landau, we assume that a can be expanded in a power series around the critical temperature T_c

$$a(T) = a'(T - T_c) \tag{6}$$

b is positive at T_c and by continuity remains positive around T_c. Equation (5) shows that $M \sim (T_c - T)^{1/2}$. At $T = T_c$, $a = 0$, and Eq. (3) shows $H \sim M^3$. From (3) we can calculate the susceptibility $\chi = dM/dH$ and get the Curie-Weiss law $\chi \sim (T - T_c)^{-1}$. The specific heat turns out to be discontinuous.

Putting $H = 0$ and $d\Phi/dM$ into (1), we get [3]

$$\frac{dM}{dt} = -D(2aM + 4bM^3) \tag{7}$$

With the initial condition $M = m$ at $t = 0$ the solution is

$$M^2 = \frac{M_0{}^2 \exp(-4aDt)}{\exp(-4aDt) - 1 + (M_0/m)^2} \tag{8}$$

where the value (5) below T_c is denoted by M_0. Above T_c, $a > 0$, M^2 diminishes to zero; below T_c, $a < 0$, and M^2 decays down to $M_0{}^2$. The long time behaviour of decay is governed by an exponential. Hence one gets a relation between the relaxation time T_1 and temperature

$$T_1(T - T_c) = \text{const} \tag{9}$$

This is the Korringa relation for relaxation. We are thus reproducing what are known as mean-field critical indices for static and dynamic responses [4, 5]. From the exact solution in two dimensions and from very accurate numerical work in three dimensions for the Ising model, we get different critical indices; thus $M_0 \sim (T_c - T)^{1/8}$, $\chi \sim |T - T_c|^{-7/4}$ in two dimensions and $M_0 \sim (T_c - T)^{5/16}$ and $\chi \sim |T - T_c|^{-5/4}$ in three dimensions. The renormalization group calculations provide similar indices. By and large the experimental work has supported these calculations. $Rb_2C_0F_4$, a two dimensional magnet, shows the magnetization index $\beta = 0.119 \pm 0.008$ and EuS, a three dimensional magnet, shows $\beta \simeq 0.33$

IV Comments

One must note that the specific critical indices of the Landau theory depend on the analyticity assumption (6). This point has been well made by Schulman [6]. The variable in the polynomial of the cusp catastrophe need to be linearly related to the most conveniently observed physical quantity—the magnetization, for instance. The most we may expect that the variable is a monotonic function of the order parameter. Without another additional assumption about this function or about the control parameters a and b, the indices cannot be calculated. The predictive powers of the catastrophe theory are rather limited.

The familiar gas-liquid transition is described rather well by the Van der Waals equation

$$(p + aV^{-2})(V - b) = RT \tag{10}$$

supplemented by the Maxwell rule. Equation (10) is a cubic equation in V, and is similar to Eq. (3). The catastrophe involved is discussed by Thom and Zeeman [1, 2]. The analogy of the gas-liquid and the magnetic case is rather subtle, and should not be pushed too far from the critical point. The van der Waals equation gives a cusp catastrophe, but it has unphysical features which are corrected by Maxwell's construction. But Maxwell's rule must be brought in as an extra assumption. It is possible that general rather than elementary catastrophes are involved, but not much seems to be known about them.

Schulman [6] and, following some of the ideas of the catastrophe theory ,Griffiths [7] have tried to analyse higher critical points. The most celebrated higher order critical point is the polycritical point of He^3 and the tricitical point of the He^3-He^4 mixtures. Experiments on fluid mixtures and metamagnets also reveal tricritical points. *Such efforts starting from the catastrophe theory are likely to be here successful because even the Landau or mean field theory produces the qualitative features of the phase diagram correctly.* Griffiths has worked out the topology of the phase diagrams of one and two order parameters in certain cases, and had tried to give generalizations of the Gibbs phase rule.

V. A Catastrophe in Nuclear Physics?

Nandy [8] has pointed out that neutron induced fission of heavy nuclei has some features of the cusp catastrophe. It is well-known that spontaneous fission is asymmetric; a nucleus of mass number around 240 breaks up into two unequal fragments of masses around 95 and 140 Qualitatively, this asymmetry is believed to be due to a fundamental asymmetry in nuclear dynamics. The nuclear forces are charge independent. However, there is Coulomb repulsion between protons but not between neutrons.

Some nuclei break up asymmetrically when a small amount of energy is supplied by a neutron absorbed by them. As the energy of the neutron increases, the fission becomes more and more symmetric until beyond a threshold there is only symmetric fission into two fragments of equal mass. The phenomenon has certainly some feature of cusp catastrophe bimodality, inaccessibility, and catastrophic split up. Analysis of the phenomenon in terms of a cusp catastrophe leads to the equation

$$(N - \bar{N})^2 = K(E - \bar{E})^3 \tag{11}$$

Here $\bar{E}$ is the Q-value or energy release in spontaneous fission. N is the neutron number of a fissioning nucleus with excitation E. $\bar{N}$ is the mean of the most probable neutron numbers at spontaneous fission. The constant of proportionality K would depend on the nature of the element.

We can use (11) as follows. Consider Fermium isotopes. Fm(256) has a two peaked distribution, while thermal neutron induced fission of Fm(255) is symmetric. Knowing the most probable neutron numbers corresponding to the A-values from fission tables and assuming the binding energy of the last neutron to 6 MeV, Nandy obtains $K \simeq 0.23$ subject to small uncertainties.

Assuming K to be constant he obtains N for Fm (254) and estimates at what neutron energy Fm (253) would give symmetric fragment distribution, He finds that with neutrons of about 1.5 MeV Fm (253) would give symmetric yield.

A more exact quantitative estimate cannot be expected. Experimental data on threshold symmetric fission are rare and only starting to accumulate. Even if the catastrophe theory succeeds in stimulating such experiments, it will have served its purpose.

REFERENCES

1. Thom, R. (1975) Structural Stability and Morphogenesis, W. Benjamin Reading, Mass.
2. Zeeman, E.C. (1977) Catastrophe Theory—Selected Papers 1972–77, Addison Wesley, Reading, Mass.
3. Majumdar, C.K. (1967) Phys. Rev. 160, 430–31.
4. Fisher, M.E. (1976) Rev. Mod. Phys. 47, 597–616.
5. Hohenberg, P.C. and Halperin, B.I. (1977) Rev. Mod. Phys. 47, 435–79.
6. Schulman, L.S. (1973) Phys. Rev. B7, 1960–67.
7. Griffiths, R.B. (1975) Phys. Rev. B12, 345–55.
8. Nandy, A. Preprint, Physics Department, Calcutta University.

EIGHT

Competitions, Games and Catastrophes

J.N. KAPUR

1. Introduction

Perhaps the three most outstanding achievements of applied mathematics of the last fifty years are: (i) the theory of ecological competitions due to Lotaka [7] and Volterra [13]; (ii) the theory of games due to Von Neumann and Morgenstern [14]; and (iii) the catastrophe theory due to Thom [12] and Zeeman [15]. Each of these has had a tremendous impact on mathematics and its applications to physical, biological, social, medical and management sciences. In the present paper, we examine the problem of competition of species from the points of view of these three theories.

2. Some Catastrophic Situations for Volterra's Competition Model

Let $N_1(t)$ and $N_2(t)$ be the populations of the two competing species at time t, then Volterra's model is given by

$$\frac{dN_1}{dt} = N_1[k_1 - b_{11} N_1 - b_{12} N_2], \quad \frac{dN_2}{dt} = N_2 [k_2 - b_{21} N_1 - b_{22} N_2] \quad (1)$$

$$\text{or } \frac{dM_1}{dt} = k_1 M_1 [1 - K_1 M_1 - K_2 M_2], \quad \frac{dN_2}{dt} = k_2 M_2 [1 - M_1 - M_2] \quad (2)$$

$$\text{where } M_1 = \frac{b_{21}}{k_2} N_1, \; M_2 = \frac{b_{22}}{k_2} N_2, \; K_1 = \frac{b_{11}}{b_{21}} \frac{k_2}{k_1}, \; K_2 = \frac{b_{12}}{b_{22}} \frac{k_2}{k_1} \quad (3)$$

There are nine possibilities according as $K_1 \gtreqless 1$, $K_2 \gtreqless 1$ (Kapur

[3]). The case $K_1 < 1$, $K_2 > 1$ is of special interest (Kapur [1]). In this case, there are two positions of stable equilibrium, viz. A: $(1/K_1, 0)$ and B: $(0,1)$ and two positions of unstable equilibrium, viz. O: $(0,0)$ and C: $[(1-K_2)/(K_1-K_2), (K_1-1)/(K_1-K_2)]$. The basins of influeuce of A and B are separated by 'separatrix' such that any trajectory starting from any point in the basin of influence of A (B) terminatns at A (B). If the initial population point (N_{10}, N_{20})

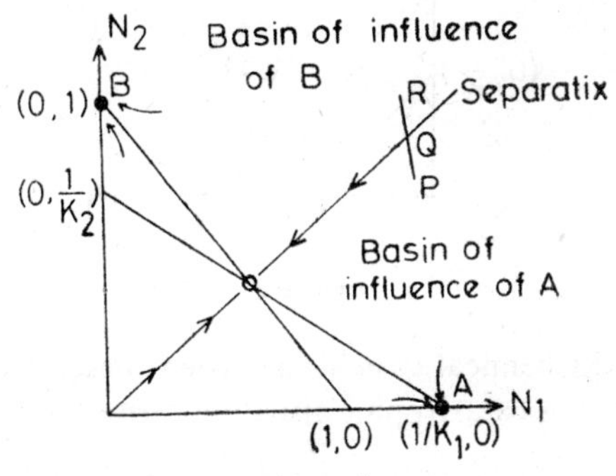

Figure 1

is at P, the second species dies out; if it is at Q, the two species coexist; and if it is at R, the first species dies out. Thus infinitesimal changes in initial population sizes can cause significant and finite or catastrophic changes in the final output of the system.

Now let us keep the initial population point fixed at Q so that the two populations ultimately coexist. If the parameters are slightly changed so that the separatrix passes through P (R), then the first (second) species dies out. Thus infinitesimal changes in parameters

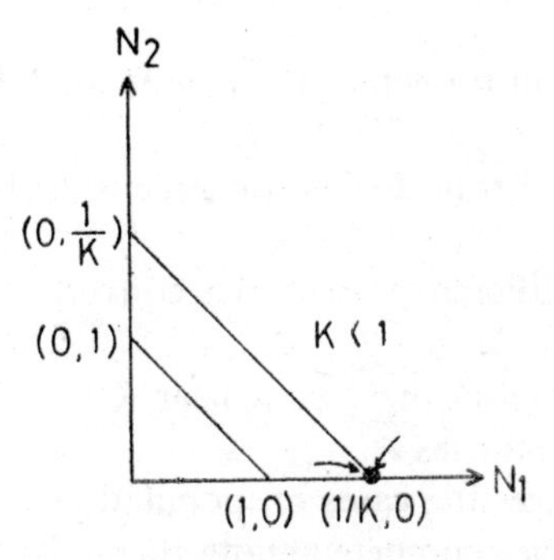

Figure 2

can cause a switch of the final population point from N_1 axis to N_2 axis or vice versa.

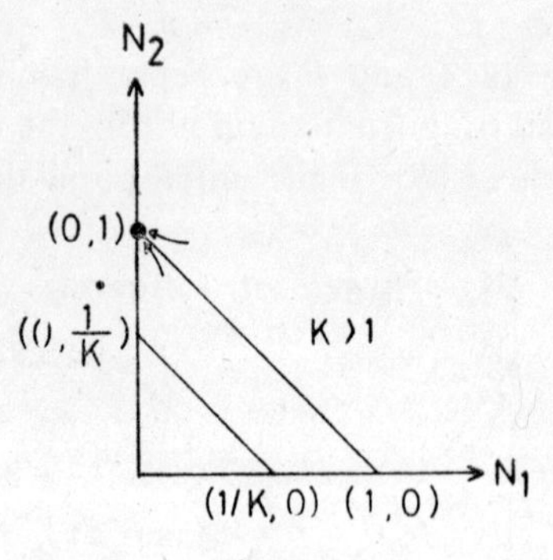

Figure 3

A similar 'biochemical switch' has been described by Rosen [8]. Another interesting catastrophic situation arises when $K_1 = K_2 = K$

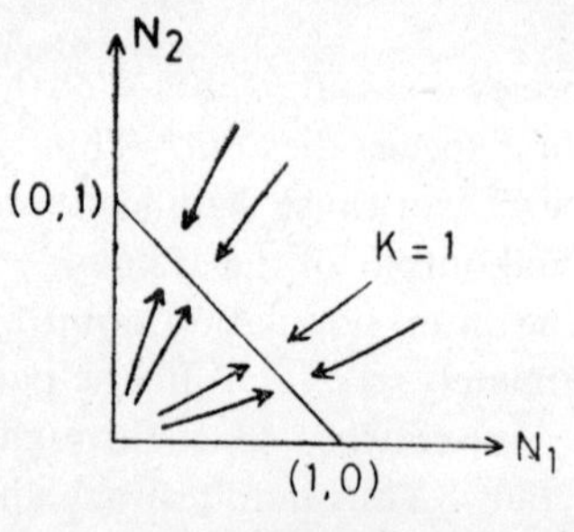

Figure 4

When $K < 1$, all trajectories converge to $(1/K, 0)$ and the second species dies out.

When $K > 1$, all trajectories converge to $(0,1)$ and the first species dies out.

When $K = 1$, different trajectories converge to different points on the line $M_1 + M_2 = 1$.

Thus infinitesimal changes in K near $K = 1$ can make significant changes in the trajectories.

Even more interesting cases of 'population switches' occur when three or more species compete among themselves.

3. Theory of Games and Population Dynamics

Maynard Smith [9,10] gave a static model for the evolution of animal conflicts by using concepts from the theory of games. Later Taylor and Jonker [11] and Zeeman [16, 17] introduced a dynamic into this model. Zeeman [16] obtained the following system of cubic differential equations

$$\frac{dx_i}{dt} = x_i \left[(\mathbf{AX})_i - (\mathbf{X}^T\mathbf{AX}) \right], \; i = 1, 2, \ldots, n \tag{4}$$

where $x_i(t)$ is the ratio of the population of the animal group following the ith strategy to the total population, $\mathbf{A}$ is the $n \times n$ pay-off matrix and $\mathbf{X}$ is the column vector $[x_1, x_2, \ldots, x_n]$.

Kapur [4] has shown that for the population model

$$\frac{dN_i}{dt} = N_i \left[k_i + \sum_{j=1}^{n} a_{ij} \frac{N_j}{N} \right]; \sum_{i=1}^{n} N_i = N; i = 1, 2, \ldots, n \tag{5}$$

for n interacting species, obtained by modifying Volterras equations, we get the system of cubic differential equations

$$\frac{x_i'}{x_i} = \sum_{j=1}^{n} \left(a_{ij} + k_i - k_j \right) x_j - \sum_{i=1}^{n} \sum_{j=1}^{n} \left(a_{ij} + k_i - k_j \right) x_i x_j \quad (i = 1, 2, \ldots, n), \tag{6}$$

where $x_i = N_i/N$. This is the same as (4) when

$$k_1 = k_2 = \ldots = k_n \tag{7}$$

When $n = 2$, (6) gives (Kapur [2])

$$x_1' = x_1 x_2 (\mathbf{A} + \mathbf{B}x_1), \; x'_2 = - x_1 x_2 (\mathbf{A} + \mathbf{B}x_1) \tag{8}$$

where

$$\mathbf{A} = k_1 - k_2 + a_{12} - a_{22}, \; \mathbf{B} = a_{11} + a_{22} - a_{12} - a_{21} \tag{9}$$

The case

$$\mathbf{A} \lhd 0, \; \mathbf{B} > 0, \; |\mathbf{A}| < |\mathbf{B}| \tag{10}$$

is specially interesting. In this case there are two stable positions of equilibrium, viz. P: (1,0) and Q: (0, 1) and one unstable position of equilibrium, viz. R:$(-A/B,\ 1+A/B)$

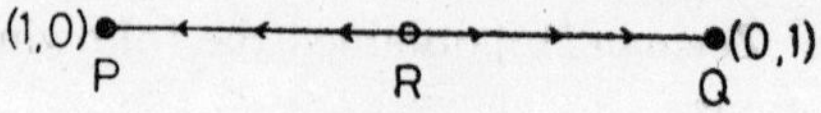

Figure 5

The flows are shown in Figure 5.

$$\text{If } x_{10} > -A/B, \text{ then } x_1 \to 1,\ x_2 \to 0 \tag{11}$$

$$\text{If } x_{10} < -A/B, \text{ then } x_1 \to 0,\ x_2 \to 1 \tag{12}$$

An infinitesimal change in x_{10} in the neighbourhood of $-A/B$ can result in a population switch from P to Q or vice versa.

'Population switching' can also be obtained by change of parameters. This also shows the phenomenon of *hysterisis*. Let

$$A_1 < 0,\ B_1 > 0,\ |A_1| < |B_1|;\ A_2 < 0,\ B_2 > 0,\ |A_2| < |B_2|,\quad |A_1/B_1| < |A_2/B_2| \tag{13}$$

then P, Q are positions of stable equilibrium and R_1, R_2 are the positions of unstable equilibrium in the two cases

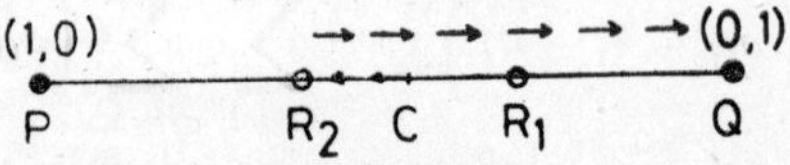

Figure 6

Let A_1, B_1 be at first the parameters and let C be the initial population point. It begins to approach P. Before it reaches R_2, let the population parameters be changed to A_2, B_2, then the population point begins to approach Q. After it has crossed R_1, let the parameters be once again changed to their original values A , B_1, then inspite of this change, the population point coverges to Q and not to P.

For three species, the basic equations are:

$$\begin{aligned} x_1' = {} & x_1x_2\,(k_1-k_2+a_{11}x_1+a_{12}x_2+a_{13}x_3-a_{21}x_1-a_{22}x_2-a_{23}x_3) \\ & + x_1x_3\,(k_1-k_3+a_{11}x_1+a_{12}x_2+a_{13}x_3-a_{31}x_1-a_{32}x_2-a_{33}x_3) \end{aligned} \tag{14}$$

$$x_2' = x_2x_3\,(k_2-k_3+a_{21}x_1+a_{22}x_2+a_{23}x_3-a_{31}x_1-a_{32}x_2-a_{33}x_3)$$
$$+\,x_2x_1\,(k_2-k_1+a_{21}x_1+a_{22}x_2+a_{23}x_3-a_{11}x_1-a_{12}x_2-a_{13}x_3) \quad (15)$$

$$x_3' = x_3x_1\,(k_3-k_1+a_{31}x_1+a_{32}x_2+a_{33}x_3-a_{11}x_1-a_{12}x_2-a_{13}x_3)$$
$$+\,x_3x_2(k_3-k_2+a_{31}x_1+a_{32}x_2+a_{33}x_3-a_{21}x_1-a_{22}x_2-a_{23}x_3) \quad (16)$$

The three positions of equilibrium (1, 0, 0), (0, 1, 0), (0, 0, 1) will be stable if

$$k_1+a_{11} > k_2+a_{21}\ ,\ k_1+a_{11} > k_3+a_{31} \quad (17)$$

$$k_2+a_{22} > k_1+a_{12}\ ,\ k_2+a_{22} > k_3+a_{32} \quad (18)$$

$$k_3+a_{33} > k_1+a_{13}\ ,\ k_3+a_{33} > k_2+a_{23} \quad (19)$$

In this case, there are three separatrices separating the three basins of influence in the 3–simplex. Infinitesimal changes in initial positions across the separatrices can cause switching of ultimate populations among these three points.

4. Bifurcations

Whether we use Volterras original equations or our form of modified Volterra's equations or the game-theoretic approach, we can find the points of equilibrium and then carry out the usual stability analysis. We get a characteristic equation in λ of degree n when we use the variables $N_i\,(t)$ and of degree $(n-1)$ when we use the variables $x_i\,(t)$. We can apply Routh-Hurwitz criteria or we can find all the real and complex roots numerically.

We are interested in finding out whether the real parts of all the roots are negative. A first step in this direction is sometimes to find the condition on the parameters that the characteristic equation may have purely imaginary roots. This has been done for two populations models by Kapur and Khan [6].

When delay effects are introduced, we get transcendental characteristic equations with an infinity of roots. The bifurcation theory for a dozen such models has been discussed by Kapur [5].

5. Concluding Remarks

Bifurcation theory, ecological competition theory, game theory and catastrophe theory were created in this order at intervals of about twenty years. Each of these has played an important role in population dynamics and the implications are still being worked out. We may expect another breakthrough, before the middle of the next decade, of great significance to population dynamics.

References

1. Kapur, J.N. (1979a) "A new catastrophe machine", IIT/K Research Report.
2. Kapur, J.N. (1979b) "A new model for population dynamics", IIT/K Research Report.
3. Kapur, J.N. (1979c) "Competition between two species", IIT/K Research Report.
4. Kapur, J.N. (1979d) "Population dynamics via games theory and modified Volterras equations", IIT/K Research Report.
5. Kapur, J.N. (1979e) "Bifurcation theory in population models with time delay", IIT/K Research Report,
6. Kapur, J.N. and Khan, Q.J.A. (1978) "Bifurcation theory for two population models", Ind. J. Pure App. Math. Vol, 9 No. 9, 787–96.
7. Lotka, A.J. (1925) Elements of Physical Biology, Williams and Wilkins, Baltimore.
8. Rosen, R. (1972). "Mechanics of epigenetic control". In R. Rosen (ed.) Foundations of Mathematical Biology Vol. 2, Academic Press, New York.
9. Maynard Smith, J. (1974) "The theory of games and evolution of animal conflicts", J. Th. Biology 41, 209–21.
10. Maynard Smith, J. (1976) "Evolution and the theory of games", Am. Scientist 64, 41–48.
11. Taylor, P.D. and Jonker, L.B. (1978) "Evolutionary stable strategies and games dynamics", Math. Biosciences 40, 145–56.
12. Thom, R. (1975) Structural Stability and Morphogeneses, Benjamins, Reading, Mass.
13. Volterra, V. (1926) "Variations and fluctuations in the numbers of individual animal species living together". In F.M. Scude and J.R. Zeigler (eds.) The Golden Age of Ecology, Springer Verlag, New York.
14. Von-Neumann, J. and Morgenstern, O. (1944) Theory of Games and Economic Behaviour, Princeton Univ. Press, Princeton.
15. Zeeman, E.C. (1977) Catastrophe Theory—Selected Papers 1972–1977, Addison Wesley, Reading, Mass.
16. Zeeman, E.C. (1979a) "Population dynamics from games theory", Proc. Int. Cong. Global Theory and Dynamical Systems, North-Western, Ill.
17. Zeeman, E.C. (1979) "Dynamics of the evolution of animal conflicts", Warwick Univ. Report.

NINE

Investigation into the Nature of Language a lagico-mathematical study

A.K. RAY

> In conclusion, we have seen that an analysis of the grammatical structures of language requires a subtle mixture of algebra, dynamics and biology. Without pretending to have a definitive answer to a problem whose difficulty can scarcely be measured, I venture to suggest that these ideas may contain something of interest for many specialists.
>
> — Rènè Thom
>
> (Structural Stability and Morphogenesis, p. 329)

INTRODUCTION

A mathematician has a good reason to take interest in the science of language, for with his mathematical methods and acumen, he is indeed in a position to make positive contribution to the subject. Only, he should also remember that language has various other functions, else than catering to the needs of mathematics and mathematical sciences. His purpose should not be to turn language itself into a procrustean bed of dead precision, but to create an awareness of its endless peculiarities, both contingent and essential. His direct purpose should be the improvement of the linguistics—the logic and

*The observations made here are my own, for whatever is their worth. I was made aware of the resemblance in spirit with Rènè Thom's theory only on the eve of the present seminar. I have couched my findings in a manner befitting the occasion.

grammar of language. It is only incidental that we mainly take up 'English' for this purpuse.

Benefits of this improved knowledge of linguistics may accrue to the main body of a language through the 'trickling down' process. For example, if the people in general become aware of the beauty and convenience of the one-one correspondence between letters and their sound values, they may one day pine for its realization.

2. Thought and Language

Language is the expression of thought and naturally, therefore analysis of language is intimately connected with analysis of thought. But the two are not the same. This is because though language expresses thought and therefore purports to be 'isomorphic' to it, the goal is ever unrealized. Language has its own peculiarities, distinct from the peculiarities of thought just as photography can have its own peculiarities apart from the peculiarities of the object of photograph.

It is suggestive to think of the possible thought total as a topological space with a sort of a rudimentary metric. Language is then a rough ϵ-net over it, of course, dynamically growing and changing. There is nothing strange about it, since any structurally stable form is a singularity in its neighbourhood and what else a word is? A word depicts a singularly realized concept from an infinite background of silent possibilities (cf. conceptualism, specially the gestalt theory; also theories of Pudgala and Vishesha in ancient Indian philosophy).

A word once created behaves very much like a living organ, trying to perpetuate its own life and adapting for that purpose to changing circumstances and suffering at times from split personalities and contradictions—a bifurcation of the ego, in the language of Catastrope Theory.

It would be interesting to consider one or two examples of the less serious and the more serious type. Consider the word 'umbrella' in English language. The word for the same object in French is 'paraplui'. Etymologically 'umbrella' means that which gives shades (and saves one from the sun) and 'paraplui' means that which saves one from the rain. But the words have braved through, in their respective habitats, the constant 'opposite' treatment ! But the sail is generally not so easy, for words of logically higher types (à la positivists). Consider, for example, the word 'logic' itself. Traditionally it means analysis of thought. How can we then countenance the uninterpreted many-valued logics? (Even for an interpreted system to deserve the appellation 'logic', the arguments shoald still be 'propositions'.) Logic which dealt with and emphasized the formal aspects of thought, was itself devoured by its own formalism. Here formalism predates over thought—a perfect piece of confusion of actants in a capture morphology. (Physicists beware, when you call your subject 'mathematical physics'. Mathematicians delighting in pure formalism and in its 'meaning-lessness' soon create inverse cube or inverse n-powered planetary systems!)

Within the scope of a single paper of this kind, it would be vain to try for any completeness in any branch of linguistics. After making two specific observations upon phonetics and spelling we shall pass over to the field of grammatical categories.

3. PHONETICS

3.1. *Composite vowels.* The vowel space for the totality of human beings is obviously a topolgoical space (that of a single race or of an individual is a subspace of this). Without entering into too much details, this space can be pictured as an open region in R^2 or R^3 as shown in the two models (dimensions corresponding to the degree of freedom of lebial movement). That we have only a few vowels in any language is easy to understand. Once a particular vowel is standardized (given a vocal habitation and a name) we

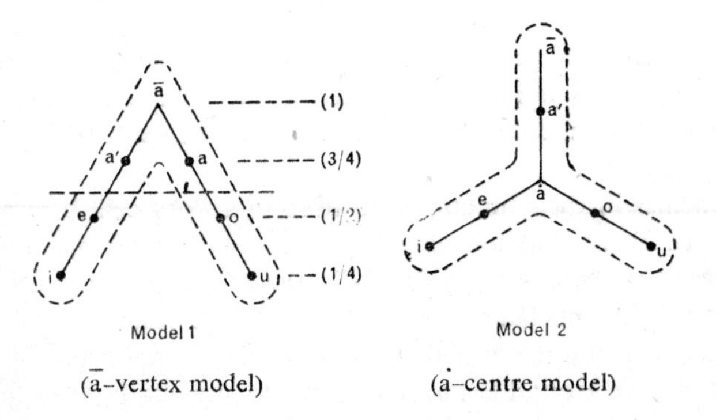

Model 1 (ā–vertex model)

Model 2 (ȧ–centre model)

must avoid another in its certain neighbourhood for fear of confusion. Let us consider the standard seven vowels as prevalent in Bengali tongue, to be denoted by **a** (ball), á (bat), ā (bar), e (bay), i (bee), o (boat), u (boat).

We norm the vowels decreasingly according to gradual closure of lips (see model 1). Thus $|\bar{a}| = 1$, $|a'| = |\dot{a}| = 3/4$, $|e| = |o| = 1/2$ and $|i| = |u| = 1/4$. We call the vowels above the midline shown, the wide vowels and those below, the narrow vowels. We are now in a position to give an adequate explanation to the occurrence of that kind of composite vowels as consist of open vowels followed by vowel closures (Yougic Swar as called in Sanskrit, not the same thing as dipthongs). Our observations are:

(i) Only the narrow vowels have the closed forms.

(ii) Any open vowel can govern the closure of a vowel of lesser or at most equal norms.

Altogether there shall be just 24 such composites. It is not necessary that we should provide separate symbols for them. But if we choose to do so, then it is an oddity, that we do so for only two of them: 'oi' and 'ou' as in Bengali or Sanskrit.

3.2 *Phonemes and Aksharas.* Phonemes are the fundamental units of speech. They are fundamental not in the sense that their structures cannot be furher analysed, but in the sense that cannot be analysed into more than one utterable parts. From structural standpoint, phonemes are primarily of the following four types,

denoted pictorially as where the segment stands for an open vowel and the small circles denote the initial or final consonants. (In further details, composite consonants can be denoted by multiple circles.)

Phonemes in their functional relation in a word-cell behaves very much like living organelles, shrinking and stretching under the ambient metabolism of the word. This fact was well recognized by the Indian phoneticians. This is clear from their definition of Akshara. It is unfortunate that an ambiguity cropped up in subsequent literature around the term due to its use both as a phonetic unit and a scriptal unit. We shall refer to the two concepts explicitly as the phonetic akshara (Dhvany-akshara) and the scriptal akshara (Lipi-akshara).

A phonetic akshara was defined as:

(i) In a word of a single open phoneme, the open phoneme is an akshara.

(ii) In a word of a single closed phoneme, the open part is an akshara.

(iii) In a word of multiple phonemes, each phoneme, except the last is an akshara, and the last phoneme is treated as in (ii).

The justification was that a single closed phoneme was much more prolonged than a single open phoneme, but in the compressed succession of phonemes in a word, the closure of a phoneme is sucked by the next phoneme.

It would be nice if each phonetic akshara could be denoted by a scriptal akshara. But a closed phoneme before the last is a single akshara. To denote it by a single composite letter would be a monstrosity. Not only there wauld be too many consonants, but these consonants would be required to be grouped into two separate bunches, interspaced by a vowel operator. The problem was ingeniously solved in two steps:

(i) Creation of composite letters for single bunches of consonants (in addition to sinlge consonant letters), these being treated with a without vowel operators (the actual practice of having a mark for a stop and none for the preferred vowel a is of no importance here).

(ii) Relaying the closure of a closed phoneme (except the last) to the next phoneme.

Thus though individually phonetic aksharas and scriptal aksharas differ, there is a perfect one-one correspondence between them in a word. Also it may be noted that if the number of phonemes in a word is n, that of aksharas (in either sense) is n or $n+1$ according as the final phoneme is open or closed.

4. Grammatical Categories

Réné Thom has wondered that the whole of human thought could be fitted into just eight grammatical categories. The question of 'whole' of human thought is out of the place here. Even at the level of any connected piece of thinking, we may require extra-sentencial categories. As regards sentencial categories the present structure seems to require modifications and further delineations. It is important to understand what distinct semantical functions a selfsame grammatical category has actually to carry out, and judge whether some further proliferation of categories is desirable. Unfortunately a thorough analysis is not possible here. Nevertheless we shall deal with certain aspects of the grammatical categories (not only of the functional but also of the autonomous types). But before that a note on substantives will be pertinent.

4.1. *A note on substantives.* Apart from nouns or pronouns, we can have phrases to denote things about which we propound. These are the substantives in a sentence. Thom has observed that in a field of rapid change, our language is ill-suited for describing the situation; this is because we fail to grasp at the true 'substantives'. In our opinion, the failure is of thought and therefore not of language whose sole purpose is to express thought. In the realm of thought the matter is profoundly important and was well known to the ancient philosophers of the east and west, and the fact is actually bourne out by the very etymology of the word substance that which stands (remains invariant) under (a process of change). The search for substantives in this sense belongs to ontology of science and metaphysics. We are concerned here only with the linguistic aspect. of the problem. (This is without any bias for or against any doctrine of linguistic autonomy.) A substantive in a sentence must refer to something that remains the same within the scope of the sentence. When we say 'Ram goes to school' it is believed that Ram remains Ram at least during the act described. The difficulty arises in any predication that connotes a change in the substantive itself. For simplicity let us confine ourselves to the case of individuals only. From logical standpoint, any change is an instance of

$$(\exists x)((x \text{ was not } \phi) \wedge (x \text{ is } \phi))$$

or equivalently, by definition,

$$(\exists x)(x \text{ becomes } \phi)$$

In actuality when x is a determinate individual, x may have a proper name α, or a unique description '$a\psi$' or 'the ψ' where ψ is again a property. (It is a deep truth of epistemology that whenever we perceive an individual, we perceive it as an individual instance of a concept. Very few of these individuals unless it is human or a human pet, is lucky enough to possess a proper name.) Now, ψ being a property, is of the same footing as ϕ, both generating classes $\{\psi\}$ and $\{\phi\}$. In language we write

(i) α becomes ϕ

(ii) the ψ becomes ϕ (similarly for '$a\psi$')
irrespective of the fact whether in (i) we would still agree to call the resultant of the change by the name α or whether in (ii) the resultant still holds the property ψ. In the latter situations in either of (i) or (ii) the substantive is to be regarded as a pseudo-substantive along the scope of the sentence and therefore a pseudo-subject of the sentence.

In actual practice, however, in case (ii), we not only describe the situation as ibid, but often describe the resultant as a '$\phi\psi$', treating ϕ as an adjective of ψ. Thus when we burn coal to cook meat, we get cooked meat and burnt coal. Though cooked meat is meat, burnt coal is not cool. Yet we regard both 'cooked' and 'burnt' as adjectives. But then an adjective is not always a qualifier. It may be what we have called a 'simulator' (see 'adjectives'). Of course, all qualifiers or simulators are not of this kind. When an adjective is acquired via transformation, we may call it a transformer adjective. These are of two kinds; agreeable and non-aggreable, say (according as the resultant agrees or disagrees with the antecedent property).

4.2. *Adjective.* It should be defined as a word syntactically adjoined to a noun acting on it semantically in various capacities.

(i) Qualifer: Examples—intelligent man, efficient headmaster, augmented matrix, spherical neighbourhood, commutative ring, etc.

[If $\phi\psi$ is a term and a $\phi\psi$ is a ψ then ϕ is a qualifier of ψ. In this case $\{\phi\psi\} \subset \{\psi\}$.]

(ii) Simulator: Examples (to be read in parity with the above set)—Neanderthal man, assistant headmaster, augmented plane, punctured neighbourhood, ternary ring, etc.

[If $\phi\,\psi$ is a term, but a $\phi\,\psi$ is not a ψ, then ϕ is a simulator on ψ (in the sense that a $\phi\,\psi$ simulates a ψ). In this case $\{\phi\,\psi\} \cap \{\psi\} = 0$.]

(iii) Numerer: Example—Ten men. (Explanation should not be necessary.)

Note. We have already talked of adjectives of the transformer type in Section 4.1, which are qualifiers or simulators.

4.3. *Conjunction.* Syntactically it is always a conjunction since it joins two sentences. Semantically it is a sentencial operator functioning as (i) logical conjunction (product), (ii) logical disjunction (sum) or (iii) implication. These may be described as the truth-functional (sentencial) operators since the truth-value of the composite sentence depends upon the values of its constituents. There are sentencial operators which are not truth-functional in this sense. Traditionally they give rise to what are called the complex sentences as distinct from the former type of sentences called the compound sentences.

Apart from this, a conjunction may act as a connective between substantives. These may be formally defined by

$$\phi(x \wedge y) = \phi(x) \wedge \phi(y) \qquad \text{def.}$$

$$\phi(x \vee y) = \phi(x) \vee \phi(y) \qquad \text{def.}$$

It may be noted that though in ordinary language such uses are rampant (Ram and Shyam, Jadu or Madhu), they have not been recognized as a separate grammatical category. If this is done, then we shall see that there are significant connectives between subtantives that have no propositional counterpart. Let

$$x \wedge y = x \text{ and } y \text{ together (as partners)}$$

Suppose $p(z)$ means 'z did this work'. Then $p(x \wedge y)$ will mean 'x and y together did this work'. It is fantastic that we parse 'together' as an adverb here. In fact 'and...together' is the connective between terms as 'if...then' is between sentences.

Of course all uses of 'and...together' need not be of this kind. Let

$$q(x, y) = x \text{ and } y \text{ together went to Bombay.}$$

It should be noted that

$$q(x, y) \text{ implies } q(x) \wedge q(y)$$

but $$q(x \wedge y) \text{ does not implies } q(x) \wedge p(y)$$

4.4. *Preposition.* The term is purely a syntactical one. In many languages (e.g. Sanskrit and its offsprings) the same function is carried by postpositons (words or suffixes). Semantically it is now-a-days called a directional term. This is not a bad description if we understand 'direction' in a wider sense and over both space and time.

The word 'by' in the sense of 'by the help of' is an exception and should be described as an instrumental term.

4.5. *Pronoun.* A pronoun in grammar is defined to be a word which is used in place of a noun. Its use is generally explained to be euphonic, to avoid repetition of the same name. It is clear that the explanation is given keeping in mind only the third person pronouns. For while 'he' ('she,' 'it') is used after at least one occurrence of the noun it replaces such is not the case with' 'I' or 'you'. (We shall not discuss the plurals.)

The fact to be recognized is that egocentric sentences form a special category of sentences. 'I' and 'you' enjoy special status in such sentences, being the egocentric terms for the speaker himself and the addressee. A third person pronoun does not have this status. We may say that while 'I' and 'you' dominate over the names of characters they denote, 'he' is dominated by the name for which it is a substitute.

There is no harm in calling them all 'pronouns' since they do the job of a noun, but it is insightful to divide them further into (i) ultranouns: pronouns of the first and the second persons and (ii) infranouns: pronouns of the third person.

4.6. *Negation.* It is altogether a separate category by itself. From logical standpoint it is a sentencial operator. In fact it is the only unary sentencial operator, others being all binary. That negation syntactically acts upon the verb is however not accidental. It has the semantic support that, verb is the central part, the nucleus, of predication (It is interesting to note the etymology of the word in this connection.) Yet, negation is not an adverb if adverbs are meant to qualify a verb. Walking fast is walking in a certain manner, but not walking is not.

4.7. *Words of emphasis.* There are certain words of emphasis that semantically act upon the whole sentence. Syntactically they act upon the verb, the reason being as given in Section 4.6. Consider, for example, the two sentences:

(i) He will surely come back.
(ii) He will quickly come back.

While 'quickly' is an adverb in the usual sense, 'surely' is not so. The word like most words of emphasis, reflects the ambient psychological state of the speaker. To parse it as an ordinary adverb seems to be dictated by false paradigms. However recalling the central place that a verb enjoys in a sentence, we may create a new category or a new subcategory of 'ambient adverbs' (treating 'ambient' as a simulator or a qualifier adjective). If the term is taken to be a subcategory then ordinary adverbs may be called intrinsic adverbs.

The use of reflexive pronouns merits a special mention here. Consider the two sentences:

(i) Ram himself went to the market
(ii) Ram takes care of himself.

The word 'himself' in (i) is for emphasis and (therefore) truth-functionally redundant (removable). The same word in (ii) is irremovable. Here (ii) is an example of what may be called a quadratic sentence where the object coincides (in denotation) with the subject. [Let $p(x, y) = x$ takes case of y. Consider: $p(x, x)$.] Precisely for this reason, however, there is no difficulty in parsing 'himself' in (ii) where it is an object. But how the word is to ba parsed in (i)? It cannot be the subject in the ordinary sense, since there is already a word for that post. It is not even 'subject in apposition' since the latter must have equal eligibility for the post. We may try 'subject in reflex' treating it as another minor category parellel to 'subject in apposition'.

4.8. *A note on subject and object.* We conclude with a brief note on these two functional categories. The notion of 'density' brought in by Réné Thom in this connection appears to be a little too simplistic. His assertion is that, the subject is more 'dense' than the object since the subject survives but the object may perish (cat eats the mouce—Thom). At this rate, however, one would hardly dare to see a star, or a mountain or even low land, for all these objects are likely to survive the mortal seer. It is the interest or the activity that seems to be the deciding factor.

TEN

Remarks on Catastrophe Theory language aspects

A.K. DAS

One of the salient features of catastrophe theory is language. In fact, Réné Thom devotes a considerable part of his treatise [1] to language vis-a-vis its biological aspects. He traces origin of language to various human responses and deals with grammar of languages in a separate appendix. He has also sought to explain the brain activity behind languages [2–6]. His notion of language stems from his main concern in catastrophe theory, namely, the hypothesis of structural stability. According to him, one can express qualitative features in terms of ordinary language. He also asserts that structural stability is inherent in language and hence, one cannot describe structurally unstable processes in terms of a simple language. The consideration of language comes up when Thom seeks to construct geometrical models of a general nature. These models are meant for explaining, as in catastrophe theory, geometry of gastrulation and other biological morphogenesis. Thom suggests that the deep structure of language is another aspect of universal morphologies. According to him a basic structure begins as a single thought represented by a bifurcation of a dynamical system describing the neurological activity. The simplest bifurcations are, what Thom calls, elementary catastrophes and these constitute basic types of spatiotemporal sentences, which in turn, form the constituents of any language. The geometric models envisaged by Thom have a universal language and it is only the ordinary language which endows them with one dimensional linear structure.

As already mentioned, use of a language is also linked up with qualitative understanding and perception. In fact, in the case of human beings language is a reflection of the ability to simulate qualitatively external processes, physical or biological. Language, as Thom sees, works as a sensory relay: if an individual A describes orally to another individual B a certain external fact or event observed by A but not observed by B, the individual B can have a qualitative understanding of what happens. This kind of qualitative (verbal) simulation is again a significant component of Thom's modelling in geometrical terms [7, 8].

Geometry has not merely a dominant position in the entire framework of catastrophe theory but also acquires a newer meaning in the context of languages and linguistics [1]. Of late, Thom has gone to the extent of distinguishing ordinary language from that of Euclidean geometry from a pedagogical standpoint [9]. It would not, therefore, be out of place to refer to some points of distinction between these two types of languages. For instance, in an ordinary language one can hardly formalize the equivalence class defined by a word (a concept) but in Euclidean geometry, the concept, a geometrical figure is formalizable. In ordinary parlance the meaning of a word is clear and so is with Euclidean geometry as both of them depend on intuition. It is in respect of the syntax that ordinary language is poor on account of the finiteness of nuclear phrases in grammar whereas it is rich for Euclidean geometry as it admits of descriptions in a wide variety of possibilities. The language of algebra which has a formal character is a step ahead of language of geometry [1]. Thom emphasizes the role of geometry, so far as syntax of language is concerned. Geometrical language gives a wider concept of syntax without taking away the element of intuition necessary for meaning. Hence the language of Euclidean geometry has the advantage of expression in "a one-dimensional combination—that of language—a morphology, a multi-dimensional structure".

In applying catastrophe theory to linguistics, one comes across three types of one dimensional catastrophe, vide, Zeeman [10]. It remains to be seen what catastrophe(s) would be in the context of geometrical linguistics and, in particular, that relating to Euclidean geometry. It would also be a worthwhile investigation if higher catastrophes emerge from considerations of basic sentences in geometry.

References

1. Thom, R. (1975) Structural Stability and Morphogenesis, W.A. Benjamin, Reading, Mass, pp. 313–15.
2. Thom, R. (1970) Topologie et Linguistque, Essays on Topology and Related Topics (ded. G, de Rham; ed. A. Haefliger and R. Narasimhan), Springer, Heidelberg, pp. 226–48.
3. Thom, R. (1973) "Language et catastrophes: Elements pour une Semantique Topologique", Dynamical Systems, Proc. Symp, Salvador, Brazil, 1971, ed. M.M. Peixoto, Academic Press, New York, pp. 619–54.
4. Thom, R. (1973) De l'icone au symbole; Esquisse d'une thèorie du symbolisme, Cahiers Internationaux de Symbolisme, 22–23, pp. 85–106.
5. Thom, R. (1973) Sur la typologie des langues naturelle: essai d'interpretation psycho-linguistique. In Formal Analysis of Natural Languages, Edition Mouton, Paris.
6. Thom, R. (1974) La' linguistique, discipline morpologique exemplaire, Critique, 322 pp. 235–45.
7. Thom, R. (1974) "A mathematical scheme for morphogenesis: structural stability and catastrophes", Methematical Models in Biology and Medicine, North-Holland, Amsterdam, pp. 93–95.
8. Thom, R. (1974) "Qualitative and quantitative models—a panel discussion", Mathematical Models in Biology and Medicine, pp. 135–136.
9. Thom, R. (1972) Modern mathematics: does it exist? Development in Mathematical Education, Proc. Second Internat. Cong. Math Edn., Cambridge University Press, pp. 194–209.
10. Zeeman, E.C. (1975) Catastrophe theory: its present state and future perspectives, linguistics. Lecture Notes in Mathematics, 468: Dynamical Systems, Warwick, 1974, ed. A Manning, Springer-Verlag, New York, pp. 381–83.

ELEVEN

Biomechanical Problems and catastrophe theory

D.K. SINHA

The study of bone constitutes one of the leading areas in biomechanics. The study of its purely mechanical aspect with an eye to its usability often brings it closer to biomedical engineering but surely brings it within the purview of mechanics of continua, with a slant to elasticity. Hence some biomechanical problems become amenable to analysis in 'Catastrophe Theory' and to 'Bifurcation theory'. Electromechanical behaviour of bone, particularly, the piezoelectric behaviour is another fascinating area of studies in bone. One of the underlying efforts in such studies has been and still is, to look for "remodelling" which is inextricably bound up wiih morphogenesis, if one is to delve deeper into it. It is here that Thom's Catastrophe Theory ought to have some relevance. The present paper is an attempt to set out perspectives in this direction with a view to applying catastrophe theory in the study of bone.

The main purpose of this paper, in keeping with the key theme of this seminar, is to look for and to dwell on areas that need to be reviewed in the context of "catastrophe theory". A dominating concern in biomechanics or strictly speaking biosolidmechanics is about bone. There is hardly any need to reiterate what it is all about. But that its use has tremendously gained in the cour se of recent years is evidenced by the proliferating literature, vide, Fukuda and Yasuda [1], Lang [2], Shamos and Lavine [3], Bassett [4], Becker and Bassett [5], some of whom have dealt with piezoelectric effects in bone. It is this effect that brings in the trail of its study, an important facet of studies and this is what has come to be known as 'remodelling', vide, Gjelsvik [6, 7]. This does also come up otherwise when we need to investigate how a bone, in a viscoelastic setting, can remodel itself in response to mechanical stress in such a way as to provide more support to the stress. Traditionally, one is used to think in terms of Wolff's law which is accounted for by mechanical energy of deformation in bone being dissipated in order to supply the necessary stimulus to initiate remodelling. What is this Wolff's law? To delve into it and look for its genesis, one has to trace back to the relationship between the form that a bone assumes and its function. According to J. Wolff [8] the law may be stated as follows: 'The form of a bone being given, the bone elements place or displace themselves in the direction of functional forces and increase their mass to reflect the amount of functional forces', (Bassett [4]). The key words, it may be noted, in this statement are 'form' and 'forces' (functional). One now requires to forge a link between 'form' and 'function' of a bone. The study of such a relationship started seriously sometime in late sixties of the last century. For 'form' one is inclined to look for a comprehensive view of the subject '*On Growth and Form*' by d' Arcy Thompson [9]. Several investigators like Murray [10], a disciple of Thompson, and Weiss [11], have discussed different aspects of bone vis a vis Wolff's law. As already mentioned, it is the 'form' of bone that dominates in these discussions. If one scans deeply this literature, one finds a quantitative approach in such studies without clarifying what 'form' is all about. We, therefore, turn to recent trends of studies about 'forms', in the context of modelling in general. We are again accustomed to the use of differential equations and more so, when we deal with buckling, deformation, etc. The study of evolution of forms assumes then a special significance. 'Form' has a variety of connotations in a variety of contexts. We can hardly think of a form without assigning its stability which is necessitated for its perception and recognition. But if forms undergo alteration, disunite, amalgamate and are affected by chaos or, in other words, morphogenesis takes place, one is faced with a peculiar situation. We could have been complacent about modelling in mathematical terms as far as it relates to forms if discontinuities do not affect the processes of evolution of forms. A morphogenesis may be represented by the equation .

$$\frac{d\mathbf{x}}{dt} = f(\mathbf{x}, \boldsymbol{\lambda})$$

$$\mathbf{x}(0) = \mathbf{x}_0$$

where $\mathbf{x}$ is n-dimensional state vector,

$\boldsymbol{\lambda}$, n-dimensional parameter vectors

f, a sufficiently smooth mapping from

$R^n \times R^n$ to R^n

A morphogenetic structure obtained from the above is given by a differential equation

$$\frac{d\mathbf{x}}{dt} = f(\mathbf{x}, \boldsymbol{\lambda}), \qquad \mathbf{x}(0) = \mathbf{x}_0$$

$$\frac{d\boldsymbol{\lambda}}{di} = g(\mathbf{x}, \boldsymbol{\lambda}), \qquad \boldsymbol{\lambda}(0) = \boldsymbol{\lambda}_0$$

where g is also a sufficiently smooth function.

If we change g further we may create another differential equation representing a form. As already said, the concept of 'form' is inextricably bound up with 'stability'. How would we ensure stability in forms? Should we look for stable forms or stable systems? How do we go about this? Possibly, stable systems may be conceived in the light of persistence of forms, in the context of perturbations. Therefore, we need to look for these systems which, in some sense, are robust against small perturbations in systems as a whole. In other words, we can say that sufficiently small changes in systems will produce behaviour which is in some sense qualitatively similar to the original behaviour of the system (see Andrew and McLone [12]). We are still bogged down with hitherto undefined terms 'sufficiently small' and 'qualitatively similar'. Can one still hope for most systems to be stable? Perhaps one can. Most systems are stable and, indeed, one has this optimism because of the theorem of Peixoto in 1962, following the earlier work of Andronov and Pontryagin that in 2-*D's the stable systems on a manifold X form an open, dense subset of the space of all systems on X, suitably topologized.* Unfortunately this does not hold good for higher dimensions; one can think about retrieving the situation by restricting to gradient like systems, i.e. systems which admit a Lyapunov function meaning a *smooth function f on X with the property that f decreases along the solution curves.* It is the shape of f that enables us to think of an approximation to study of stable systems and we are then led to stable functions: A function $f: x \rightarrow R$ is said to be stable if every close function $f': x \rightarrow R$ is equivalent to f in that there exist diffeomorphisms $h: X \rightarrow X$ and $k: R \rightarrow R$ such that

$$X \xrightarrow{h} X \xrightarrow{f'} R \xrightarrow{k} R$$

is equal to f so that the transformations h and k convert f' back into f.

Coming to brasstacks, our immediate interest is to study equilibrium states of dynamical systems which essentially imply a study of the *local variant* leading to the notion of *local stability.* Hence, we have to have a locally stable function. A function f is *locally stable* if any sufficiently close function f is everywhere locally equivalent to f.

Most functions are locally stable; for, they are those where critical (stationary) points are nondegenerate and these functions form an open dense subset of the suitably topologized space $f(X)$ of all smooth functions on X. This is how we are forced to reckon with catastrophe theory in which we are concerned with description of the ways in which equilibrium states of a dynamical system can change or bifurcate as the system varies with respect to a parameter. It is in the context of gradient like systems we now see that we have to study families $f_c: X \rightarrow R$ of functions or equivalently $f: X \times C \rightarrow R$ where C is some k-dimensional parameter space. We are then immediately led to what Zeeman [13] has done in regard to qualitative behaviour of a wide class of differential equations which need not be repeated. Thus, summing up, we can say that study of bone leads to, motivates and necessitates the study of catastrophe theory, if 'remodelling' and 'evolution' of forms are to be probed in depth, qualitatively and quantitatively.

References

1. Fukuda, E. and Yasuda, I. (1957) Phys. Soc. Jap. 12, 1158.
2. Lang, S.B. (1969) Nature, 224, 798.

3. Shamos, M.H. and Lavine, L.S. (1967) Nature, 213, 267.
4. Bassett, C.A.L. (1971) Physiological Basis of Rehabilitation Medicine, W.B. Saunders, New York, 28, 3.
5. Bassett, C.A.L. and Beeker, R.O. (1962) Science, 137, 1063.
6. Gjelsvik, A. (1973) J. Biomech. 6, 69.
7. Gjelsvik, A. (1973) J. Biomech. 6, 187,
8. Wolff, J. (1892) Das Gesetz der Transformation der Knochen, Hirschwald, Berlin.
9. Thompson, D' Arey (1942) On Growth and Form, 2nd ed. Cambridge University Press, Cambridge.
10. Murray, P.D.F. (1936) Bones: A Study of the Development and Structure of the Vertebrate Skeleton, Cambridge University Press, Cambridge.
11. Weiss, P.A. (1965) "From cell dynamics to tissue architecture," In Advanced Study Institute on Structure and Function of Connective and Skeletal Tissue, Butterworths, London.
12. Andrews, J.G. and McLone, R.R. (1976) Mathematical Modelling, Butterworths, London.
13. Zeeman, E.C. (1978) Catastrophe Theory and Its Applications, Addison Wesley, Readings, Mass.

TWELVE

Catastrophe Theory in Some Aspects of earth sciences

S. MITRA
D.K.SINHA

INTRODUCTION

As in any physical system catastrophe theory has wide applications in geological phenomena. Of the most evident ones mention may be made of earthquakes, faulting and buckling of rocks, shock wave propagation in low velocity layers, etc.; less evident cases may be geomagnetic reversal, convective heat flow (Benard), phase transitions in the earth's mantle, wobbling of the earth's rotation, and so on.

So far, however, not many have worked on geological problems. Henley's work [3] falls directly into this category but there exists a host of others which indirectly come within its purview through other sciences with which earth sciences have a strong relevance.

The purpose of the present work is to present a broad synoptic overview of such studies and bringing out the applicability of this theory in such phenomena as damping of seismic waves in the asthenosphere, etc. At the outset Henley's contribution has duly been brought into the perspective of our discussion, while at a later stage we draw upon areas of studies in other sciences, particularly in structural and engineering sciences, that have a strong bearing on problems of structural geology. To confine the scope to physical systems we have left out the analysis related to palaeontological problems. Although the works of various catastrophe theorists have been consulted concepts have largely been drawn out of Zeeman's to put to work in geo-models as may be evidenced in the text. The work is wound up with remarks pointing to directions in which studies have scope to proceed.

Structural Geology

Following the ideas set forth by Thom [1] and Zeeman [2] one becomes accustomed to Catastrophe Theory in which there is a dynamical system characterized by variables and parameters. They have made inroads into applications in physical, social and biological sciences. A continuous input of parameters brings about discontinuities or discontinuous outputs called catastrophes.

Temperature and pressure control density and gas-liquid phase transition constitute a cusp catastrophe (Fig. 1). Gas-liquid continuity may be observed by going round the cusp. Within the cusp the catastrophes exist. Both the catastrophes, however, can be delayed in the metastable states of superheated liquid or gas. It is known that "clear water at atmospheric pressure can be gently heated beyond 200 °C before boiling and when it does boil it explodes catastrophically with a sound." The thunderous sound accompanying the volcanic outburst is not as much due to the blowing off of the solidified lava on the volcanic mouth as it is due to the catastrophically bursting transition in the metastable state. Indeed, this principle of metastable states are employed in the bubble and cloud chambers.

Transformation of a foam (liquid with gas bubbles) to a spray (a gas with liquid particles) is a common example of catastrophic jump. The gravimetric gas-liquid ratio and the confining pressure constitute the two *control variables* while viscosity is the *response variable*. A magma at some depth may well afford a situation for a liquid containing gas bubbles under high pressure and having high viscosity, The ascent of magma causes a decrease of pressure which, in turn, brings about an increase in the volume of the contained gas, till the gas bubbles coalesce and then the magma is transformed suddenly from a spray back to a foam and this requires either loss of gas or increase of pressure. This results into a field where either a foam or a spray is stable. One can think of cusp catastrophe in such a situation.

The *origin of turbidites* may involve another geological situation for a catastrophe. The point of interest being the rapid transition from laminar flow to turbulence. The velocity gradient above the water-sediment interface and the viscosity of the water (varying with the content of the suspended sediment) may be taken as two control variables. The degree of turbulence may be treated as a response variable.

Fault movement is another example where one can apply qualitative modelling (on the basis of catastrophe theory) of fault movement which normally takes place after long periods of quiescence. The friction and shearing energy are the two *control varaibles* and the rate of movements as the response variable. This catastrophe is, again, a cusp one.

Henley [3] mentions geomorphological processes (e.g. river capture), geochemical and mineralogical reactions and phenomena in structural geology, e.g. Kink-banding, cleavage development, etc. in which discontinuities are significant features.

Many problems in geophysics may be treated with this theory We attempt here to deal with one problem related to the asthenosphere.

To account for the damping of the elastic waves at the low-velocity zone an assumption is made (Mitra, in press) that the ferromagnesian silicates present there contain lattice-defects, both point-defects and dislocations.

A large amount of seismic waves pass through the low-velocity layer and bring about changes in the arrangement of dislocation pinning points (impurity as well as dislocation network intersection can pin-down dislocations), which alter the dislocation density as well as the loop length. The dislocation segments between pinning points bow out in phase with applied oscillatory stress. The dynamics of dislocation motion depends on the intersection of dislocation with point defects, phonons, electrons and other dislocations in the lattice.

The sudden damping of the vibration of these dislocations in the asthenosphere can be modelled using the Duffing's equation of forced oscillations as:

$$\ddot{x} + k\dot{x} + x + \alpha x^3 = F \cos \Omega t$$

Neglecting the fourth term since α is a small nonlinar term, the equation becomes

$$\ddot{x} + k\dot{x} + x = F \cos \Omega t$$

where $F \cos \Omega t$ is a small periodic forcing term with frequency Ω close to 1, the frequency of the linear oscillator and k is a small positive damping term. The amplitude of the resulting oscillation depends on the parameters.

There are two cusp catastrophes with α, Ω as *conflicting factors*. At each cusp the upper and lower sheets represent attractors (stable periodic solutions) while the middle sheet represents saddles (unstable periodic solutions). If the frequency of the forcing term is gradually changed so as to cross one of the cusp lines, going from the inside to the outside of the cusp, then the amplitude will exhibit a catastrophic jump. There will be a sudden phase-shift at the same time. This is what may be responsible for the S wave changes.

Catastrophic events also occur in the course of formation and evolution of mixed layers in the surface waters of the ocean. One comes across turbulent missing processes. The identification of parameters is difficult.

There are many situations in structural geology which have a strong qualitative flavour, largely because of the intrinsically mechanical character of problems concerned. Indeed, it is in respect of consideration of mechanical situation or situations in mechanics that one really talks about 'morphogenesis' in a mechanical setting. Also there do occur geotectonic situations which often involve essentially plate-vibrations, breaking of plates, investigation of artificial load, etc. We consider now some examples which we draw upon from Zeeman [11].

Example 1 The simple Euler arch.
This is a simple example consisting of two rigid arms supported at the ends and pivoted at the centre, with a spring tending to keep them at 180°, as illustrated in Fig. 1.

If the ends are compressed with a gradually increasing horizontal force β then the arms will remain horizontal until β reaches a critical value, when they will begin to buckle upward (or downward). If β is now fixed, and a gradually increasing vertical load α is applied to the first, as in Fig. 2, then the arch will support the load until α reaches a critical value, when it will snap catastrophically into the downwards position. It is this behaviour that will be explained by our first cusp catastrophe.

Let the arms each have length l, and μ denote the modulus of elasticity of the spring. Initially let $\alpha = 0$. It can be proved that the arch buckles when $\beta = 2\mu$. Let $\beta = 2\mu + b$, and let x denote the angle of the arms to the horizontal. We assume a, b, x are small. In

3-dimensions let us choose the (α, β)-axes to be horizontal, and the x-axis vertical. Let the horizontal (α, β) plane be called the control plane C. Let M be the graph of X as a function of α, β.

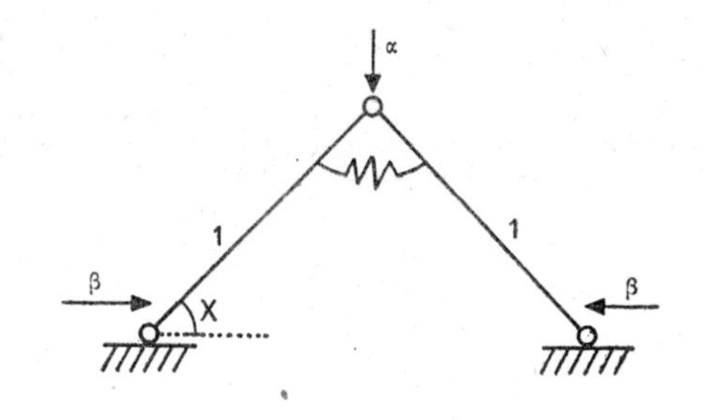

Figure 1.

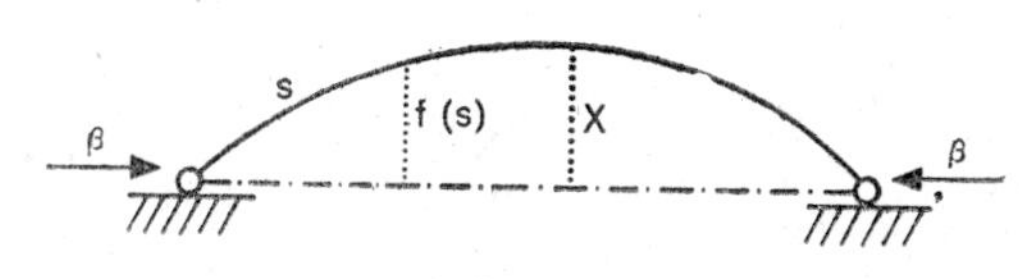

Figure 2.

It can be proved that M is a cusp catastrophe with $(-\alpha)$ as normal factor and β as splitting factor.

The proofs are in Zeeman [11].

Example 2. *The Euler strut.* We now turn from the discrete to the continuous, from the simple pivot to the elastic strut: compressed under force β, as shown in Fig. 2.

Let λ denote the length of the strut and μ the mod of elasticity per unit length.

It can be shown that M is a cusp catastrophe with $(-\alpha)$ as normal factor and β as splitting factor.

In other words the Euler strut behaves exactly as the simple Euler arch in the previous example.

Dodson and Dodson [6] undertook the analysis of both continuous and discrete valued systems in which case, the potential unction for the continuous dynamical case is given by

$$V(T_0, F, x) = K\lambda\mu \left\{ \frac{1}{4} x^3 + \frac{1}{2\lambda\mu} (\beta + E) x^2 - \frac{F}{k\lambda\mu} x \right\},$$

where T_0 is rest tension, F the force per unit length, x displacement, and the potential function, for the discrete case, is

$$V(T_0, F, x) = \frac{\lambda}{a^2}\left\{ \frac{1}{4}x^4 + \frac{1}{\lambda}(T_0 + E)\, x^2 - \frac{F\, a^2}{\lambda^2} \right\}$$

This catastrophe sets in which discontinuities occur in the equation are given by

$$4\, k^2 (T_0 + E)^2 + 27\, \lambda\, \mu\, F^2 = 0 \quad . \tag{1}$$

which is also a semicubical parabola and

$$32\, (T_0 + E)^2 + 27\, a\, \lambda\, \mu\, F^2 = 0 \tag{2}$$

which is a semicubical parabola. One can now proceed to investigate stability of the systems.

We also refer to papers by Chillingworth [10] and Sewell [9] in this connection.

Concluding Remarks

While one can think of a variety of situations in earth sciences that may be phenomenologically amenable to catastrophe theory, the identification of the nature of the catastrophe is not an easy task; for this it is necessary to get at the form of the potential function with variables and parameters with appropriate stipulation. Studies on stability, which have hitherto been a feature of studies in catastrophe theory applied to structural systems, have not made strides in regard to problems in earth sciences and this is, therefore, one of the areas that ought to develop in the years to come. As shown in some of the examples, it is through applications of catastrophe studies, one can aim at qualitative modelling of problems in earth sciences. It would be worthwhile if discontinuities observed in geological studies really compared with "elementary catastrophes" so far identified by Thom; such studies might lead to new kinds of catastrophes. Bifurcation theory, an adjunct of catastrophe theory (Thompson [10]) is yet to find application in earth sciences.

References

1. Thom, R. (1968) Une. theorie dynamique de la morphogenese, Towards Theoretical Biology, 1, 152.
2. Zeeman, E.C. Applications of Catastrophe Theory, Int. Conf. on Manifolds, Tokyo Univ. Press, Tokyo, pp. 11–23.
3. Henley, S. (1976) Catastrophe Theory Models in Geology. Math. Geol. 8 (6), 649–55.
4. Thom, R. (1972) Stabilite Structurelle et Morphogenese, Benjamin, New York.
5. Zeeman, E.C. (1976) Scientific American, 234, 65–83.
6. Dodson, C.T.J. and Dodson, M.N. (1974) York University preprint,
7. Andrews, J.G. and McLone, R.R. (1976) Mathematical Modelling, Butterworths, London.
8. Sewell, M.J. (1976) Bull. Inst. Math. Appl. 12, 163–72.
9. Thomson, J.M.T. (1975) Nature, 254.
10. Chillingworth, D. (1974) "The catastrophe of a buckling beam", In A. Manning (ed.) Dynamical Systems, Lecture Notes in Mathematics 468, Springer, Berlin.
11. Zeeman, E.C. (1977) Catastrophe Theory and Its Applications, Addison Wesley, Readings, Mass.

THIRTEEN

Catastrophes and Bifurcations in biochemical networks

A.B. ROY

The emergence of multiple steady states and bifurcation of time periodic solutions are important in the study of regulatory biochemical networks. The situation where the system can exhibit more than one simultaneously stable steady state is of particular interest. In this case a certain functional order is possible through transitions between these steady states and as a result there is a self-regulation of the concentrations of the various chemicals. In situations, involving bifurcation of time periodic solutions, there is a possibility of cooperative behaviour in the form of a temporal organization in the system. From temporal organization one gets the idea of structural stability. Two periodic (i.e. closed) trajectories in the phase space due to bifurcation mechanism are necessarily separated by a finite distance from each other as well as from the singular point. These periodic trajectories are called limit cycles. More complicated bifurcations of limit cycles are possible in the presence of separatrice loops joining two singular points. These aspects are dealt in details by Andronov, Vit and Khakin (1966).

In the present paper, we shall consider two such situations where the equations of evolution admit more than one steady state solutions.. This leads to abrupt transition between simultaneously stable steady states. This result is expressed in the language of catastrophe theory. Lastly we shall consider control circuits in biochemical pathways, which provide small amplitude periodic solutions.

INDUCIBLE SYSTEM WITH MULTIPLE STEADY STATES

(i) In Jacob and Monod's model (1961), it is observed that each structural gene that codes for an enzyme or protein is connected with an operator gene acting as a regulator for initiation of transcription. Transcription is blocked when a repressor is bound to the operator gene. In the case of inducible enzymes the binding of repressor molecule exists in the absence of effector and transcription is blocked. When the substrate for such an enzyme is present, an effector molecule can bind with repressor and thereby depressors the binding to the operator. The dynamic behaviour of a sequence of reactions under allosteric control can be deduced for systems controlled at the enzyme synthesis level and corresponding kinetic equations (for single feedback control loop) can be written in the following dimensionless form (Tyson and Othmer, 1978).

$$\left.\begin{aligned} \frac{dx_1}{dt} &= f(x_n) - K_1 x_1 \\ \frac{dx_j}{dt} &= x_j - 1 - K_j x_j \end{aligned}\right\}, \ 2 \leqslant j < h \tag{1}$$

where $x_i = a_i S_i$, S_i = concentration of i th substrate,

$$f(x_n) = \frac{1 + x_n^p}{K + x_n^p}, \quad K = 1 + K_2 R_t \ (>1)$$

$$K_2 = \frac{[OR]}{[O][R]}, \ [OR] = \text{concentration of } OR \text{ complex}$$

$$[R] = \text{concentration of repressor}$$

$$[O] = \text{concentration of operator}$$

$$p = \text{Hill coefficient}, \quad K_j = \frac{k_j}{b}$$

The steady state solution of (1) is the vector $\mathbf{x}^* = (x_1^*, x_2^*, \ldots, x_n^*)$

where $$f(x_n^*) = \phi x_n^*, \quad \phi = K_1 K_2 \ldots K_n \tag{2}$$

and $$x_1^* = K_2 x_2^* = K_2 K_3 x_3^* = \ldots = K_2 K_3 \ldots K_n x_n^*$$

The system (1) may possess either one or three solutions x_n^* of (2). The combined equation gives

$$x_n^{2p} - [K(p-1) - (p+1)]\, x_n^p + K = 0 \tag{3}$$

Equation (3) has positive real solution x_n for $K > 1$ and $p \geqslant 1$ provided

$$K > K_{\min} = \left(\frac{p+1}{p-1}\right)^2 \tag{4}$$

In the case $p = 2$, $K \gg 1$ the steady states satisfy

$$K_n^3 - ax_n^2 + Kx_n - a_n = 0 \tag{5}$$

where $a = \phi^{-1}$, ϕ being inversely proportional to S_0

For $a \ll K$, Equation (5) has one and only one positive real root and for $a \gg K$ there is also only one positive root. But if $2\sqrt{K} < a < K/2$, Equation (5) possesses three positive real roots. So the system possesses multiple steady states. Steady state diagram representing concentration X_n versus the parameter a displays multiple steady states hysteresis. The region of multiple steady states buffers the system against dramatic fluctuations in enzyme concentrations of S_0 between two extremes. This type of situation leads to abrupt transitions between simultaneously stable steady states and give rise to some interesting thermodynamic problem (Nicolis and Prigogine, 1977). This result can be expressed in the language of cusp-catastrophe and in the next model it is expressed in such language.

(ii) We shall now consider a pharmacode dynamic model as developed by Dutta and Roy (1979) to explain drug response if the drug is introduced repeatedly in an organism. In framing the model it is assumed that the drug D at the biophase level first birds with target T following the law of mass action and drug target complex (DT), so formed acts as an inducer to depress the binding of repressor to the operator region. This control mechanism is analogous to that of operon model of Jacob and Monod. This type of depression model was proposed by Goldstein and Goldstein (1961), Suster (1961) and Collier (1965) to understand the gradual attenuation of original responses of the drug for long term treatment. The kinetic equation governing the behaviour of the control biochemical network can be written as

$$\dot{x} = R(y) - k_1 x\,(d - y) + k_{-1}\, y - g(x) \tag{6}$$

$$\dot{y} = k_1 x(d - y) - k_{-1}\, y \tag{7}$$

where $R(Y)$ is a sigmoidel function of y

$$y = [DT]$$

$$x = [T]$$

$$g(x) = \text{degradation rate of } x$$

$$d = [D] + [DT] = \text{drug concentration at the biophase } (= \text{constant})$$

Choosing $$R(y) = \alpha_1 \frac{1 + K_1 y^p}{K + K_1 y^p}, \qquad g(x) = k_{10} x$$

where α_1 = constant of production, k_{10} = removal rate constant,

k_1 = measure of tightness of the binding of $[DT]$ complex to represssor.

Equations (6) and (7) can be written in the following dimensionless form:

$$\frac{dX}{d\tau} = \frac{1+Y^p}{K+Y^p} - (a_1 + a_3) X + a_2 XY + a_4 Y \tag{8}$$

$$\frac{dY}{d\tau} = \frac{a_1}{a_3} Y - XY - \frac{a_4}{a_3} Y \tag{9}$$

where $X = \sqrt{\frac{k_1}{\alpha_1}} x, \quad Y = (K_1)^{1/p} y, \quad \tau = \sqrt{k_1 \alpha_1}\, t$

$$a_1 = \sqrt{\frac{K_1}{\alpha_1}}\, d,\ a_2 = \frac{k_{10}}{\sqrt{k_1 \alpha_1}}, \quad a_3 = \sqrt{\frac{k_1}{\alpha_1}}\, \frac{1}{(K_1)^{1/p}}$$

$$a_4 = \frac{k-1}{\alpha_1} \left(\frac{1}{K_1}\right)^{1/p}$$

Here all a_i's are positive and a_1 only depends on 'd'.

The above system possesses a single steady state for $p = 1$. But if $p = 2$ the steady state is $[a_4 Y_0/(a_1 - a_3 Y_0),\ Y_0]$ where Y_0 obeys the equation

$$Y^3 - \alpha Y^2 + \beta Y - \alpha = 0 \tag{10}$$

where $$\alpha = \frac{k_1 \alpha_1 (K_1)^{1/p}\, d}{k_1 \alpha_1 + k_{-1} k_{10}} > 0, \quad \beta = \frac{K k_{10} k_{-1} + k_1 \alpha_1}{k_1 \alpha_1 + k_{-1} k_{10}}$$

It is easily seen that for $\alpha \ll \beta$ and also for $\alpha \gg \beta$, i.e. if d is either small enough compared to K or large enough compared to K,

Equation (10) has only one positive root. If $2\sqrt{\beta} < \alpha < \beta/2$ then Equation (10) admits three positive real roots. So the given systems (8) and (9) possess three steady states provided α and β satisfy the above equality and this inequality separates the parameter space into regions of single steady state and multiple steady states. If expresssed in original parameters the above inequality provides the limits of the concentration of drug '*d*' at the biophase which can provide multiple steady states. The steady state values of x and y for a given k and d are obtained from the cubic (10) with the help of computer and the stability analysis for such steady states are made by normal mode analysis. It is found that steady states with higher and lower values of y are stable and with intermediate values of y they are unstable (saddle point).

The above result can be phrased in the language of catastrophe theory. Setting

$$Z = Y - \alpha/3$$

we obtain a cubic equation from (10) in the form

$$Z^3 + uZ + v = 0 \tag{11}$$

where
$$u = \frac{3\beta - \alpha^2}{3} = u(\alpha, \beta)$$

$$v = \frac{\alpha^3 - 27\alpha}{27} = v(\alpha)$$

Solving Equations (11) for all pairs (u, v) traces out a surface in (Z, a, b) space, the catastrophe manifold, generated by the loci of maximum and minimum of the associated potential energy function $V(u, v)$. Changes in the magnitudes of the control parameters u and v may cause the system to assume diffferent stationary states. The values of the controlling prameters u and v for which one or other of the stationary states disappear may be obtained from the potential function (Thom 1972, Zeeman 1976, Woodcock 1977):

$$V(u, v) = \frac{Z^4}{4} + u\frac{Z^2}{2} + vZ$$

The equation

$$4u^3 + 27v^2 = 0 \tag{12}$$

describes the bifurcation set of values of (u, v) at which two of the stationary states vanish as two of the roots of the Equation (11) become complex.

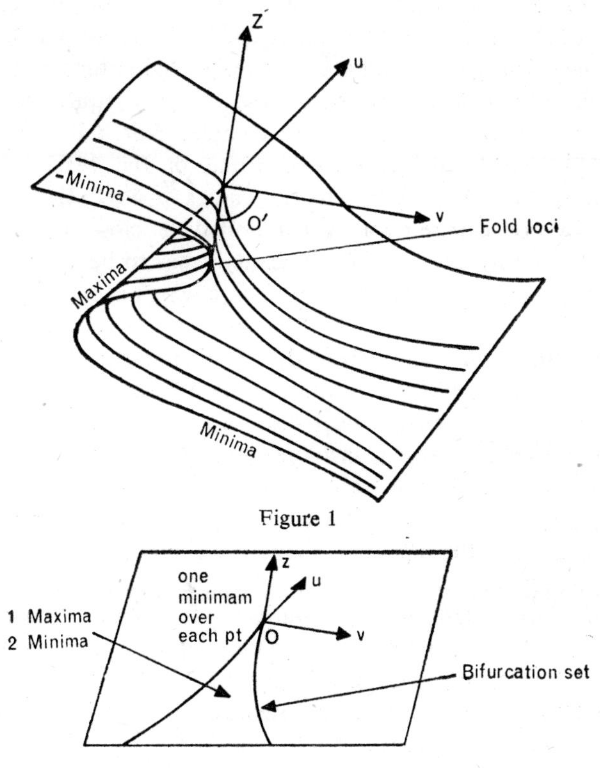

Figure 1

Figure 2

The associated manifold and bifurcation set of the simple cusp catastrophe are shown in Figs. 1, 2 respectively.

In the space of phyrico-chemical parameters α and β the behaviour is qualitatively similar. From above it is clear that it is not always possible to construct a potential function generating the equations of evolution.

Bifurcation Theory Applied to a Simple Model of a Biochemical Oscillator

Oscillations in biochemical systems are most intriguing of all chemical oscillations. At the genetic (e.g. enzyme synthesis) and metabolic levels there are ample evidences of such phenomena. To take into account such oscillatory phenomena analytically the Hopf bifurcation is a very useful tool as it refers to the development of periodic orbits (self-oscillation) from a stable fixed point, as a parameter of the system crosses a critical value. The essence of Hopf bifurcation of time periodic solution can be summarized as follows (Macdonald 1977). Consider a set of ordinary differential equations with equilibrium point at which the roots of the stability equation all have negative real parts except for the complex pair. If the complex eigenvalues cross the imaginary axis as one of the real parameters, say μ, varies through certain critical values $\mu = \mu_c$, then for near critical values of μ there exist critical limit cycle close to the equilibrium point. Just how near to criticality μ has to be is not known and unless a certain complicated expression which is called 'the magic number' (Mess and Rapp 1978), existence is only guaranteed exactly at criticality. The sign of the 'magic number' determines the stability of the limit cycle, and whether the limit cycle exists for subcritical ($\mu < \mu_c$) or supercritical ($\mu > \mu_c$). Subcritical bifurcation means that there is a closed unstable trajectory from which trajectories spiral in towards the stable equilibrium point and supercritical bifurcation means that there is a closed stable trajectory with spiral trajectories leaving the unstable equilibrium point and approaching this closed trajectory.

A general treatment of Hopf bifurcation is given in Marsden and MacCracken (1976) but Poore's (1975) method is found to be more convenient and useful in application.

We shall mention here just one situation where Hopf bifurcation theory is applied to establish stable periodic solution. The model under consideration is taken as 'Goodwin oscillator' in the general form as given by Marales and McKay (1967). The model involves a sequence of biochemical reactions with end point inhibition. Macdonald (1977) applied the Poore's method to establish the existence of small periodic solution around the unstable equilibrium point for the model described by the equations (Macdonald 1977):

$$\left.\begin{aligned} \frac{dx_1}{dt} &= \frac{c_0}{1 + \alpha x_n{}^p} - k_1 x_1 \\ \frac{dx_i}{dt} &= h_i x_{i-1} - k_i x_i \end{aligned}\right\} \quad i = 2, 3, \ldots, n$$

for a sequence of biochemical reactions of which the first involves end product intrubution and the rest follows first order kinetics.

Macdonald (1977) and Tyson and Othmer (1978) have made exhaustive study of control circuit in biochemical pathways and small amplitude periodic solutions particularly with the help of Hopf bifurcation method.

REFERENCES

1. Andronov, A.A, Vit, A.A. and Khaikin, C.E. (1966) Theory of Oscillations, Pergamon, Oxford.
2. Coolier, H.O.J. (1965) "A general theory of genesis of drug dependence of inhibition of receptor", Nature, 205, 181.
3. Dutta, B. and Roy, A.B. (1979) Proceedings of the Second Int. Conf. in Math. Modeling, (in press), St. Louis, Missouri, Rolla.
4. Goldstein, D.B. and Goldstein, A. (1961) "Possible role of enzyme inhibition and repression in drug tolerance and addiction", Biochem. Pharmacol., 8, 18.
5. Jacob, F. and Monod, J. (1961) J. Mol. Biol., 52, 343.
6. Macdonald, N. (1978) Lecture Notes in Biomathematics, 27, Springer, Berlin.
7. Marsden, J. and McCracken, M. (1976) The Hopf Bifurcation and Its Application, Lectures in Applied Mathematics, Springer, Berlin.
8. Poore, A.B. (1975) "Bifurcation of periodic orbits in a chemical reaction problem", Math. Biosciences, 26, 99,
9. Nicolis, G. and Prigogine, I. (1977) Self Organization in Non-equilibrium Systems, Acadimic Press, New York.
10. Suster, L. (1961) "Repression and derepression of enzyme synthesis as a possible explanation of some aspects of drug action", Nature, 189, 314.
11. Thom, R. (1972) Stabilite Structurelle et Morphogenesis, Benjamin, Reading.
12. Tyson, T. and Othmer, H.G. (1978) The dynamics of control circuits in biochemical pathways. In *R*. Rosen and *F.N*. Snell (eds.) Progress in Theoretical Biol., Academic Press, New York.
13. Woodcock, A.E.R. (1977) Lecture notes in Biomatnematics, 18, 343.
14. Zeeman, E.C. (1976) "Catastrophe theory", Scientific American, 234, 65.

FOURTEEN

Bifurcation Theory, Applications in physical and biophysical systems

DILIP SEN

1. Introduction

The purpose of this paper is to give an idea, with simple illustrations, how bifurcation phenomena are extremely useful for the study of physical and biophysical systems. Various aspects of bifurcation theory have been discussed in a number of books, [1–5] where mostly bifurcation of fixed points have been considered, i.e. attention confined to the bifurcation of equilibrium states. For some recent treatment of dynamic bifurcation theory one may study [6].

One of the tantalizing problems in physical, biophysical and sociological systems is the emergence of new patterns and it is imperative that we try to understand the basic aspects underlying self-organization. The evolution equations representing these systems of interest to us are in general nonlinear so that the natural approach to the problem of the emergence of new patterns or a new phase is in terms of bifurcation theory—where one studies the stability aspects of the solution space. One actually studies the possible branching of solutions that may arise under certain variations of conditions specified in terms of a set of parameters (control parameters). The branching of solutions in a nonlinear system leads to a gradual acquisition of autonomy from the environment.

2. Basic Framework

We write the evolution equation as

$$\frac{dX}{dt} = f(X, \lambda) \tag{1}$$

where

$$f : X \times \mathcal{R}^r \rightarrow Y \tag{2}$$

is a given C^k mapping ($k \geqslant 2$) and $X \subset Y$ are manifolds and $\mathcal{R}^r$ is the parameter space. Equation (1) defines a semi-flow

$$F_t^\lambda : X \rightarrow X$$

if we let $F_t^\lambda(X_0)$ be the solution of $\dot{X} = f(X, \lambda)$ with initial condition $X(\theta) = X_0$. A *fixed point* is a point (X_0, λ) such that $f(X_0, \lambda) = 0$. Then $F_t^\lambda(X_0) = X_0$, i.e. X_0 is on equilibrium point of the dynamics. Now let $X_0(\lambda)$ be a given γ-parameter manifold of solutions of $f(X, \lambda) = 0$, i.e. $f(X_0(\lambda), \lambda) = 0$ for λ in an open set in $\mathcal{R}^r$. Then (X_0, λ_0) is a bifurcation point if every neighbourhood of (X_0, λ_0) contains a solution (X, λ) of $f(X, \lambda) = 0$ with $X \neq X_0(\lambda)$. The set of all solutions near (X_0, λ_0), including $(X_0(\lambda), \lambda)$, constitute the bifurcation set. Generally, we may say that near a bifurcation point the solutions change topological type as λ is varied.

To illustrate we first consider a *linear eigenvalue problem*

$$LX = \lambda X \tag{3}$$

where λ is a real number and L a linear operator acting on vectors X in some normed linear space. For every value of λ a solution of (3) is

$$X = 0 \tag{4}$$

If we assume that there is a sequence of eigenvalues $\lambda_1 < \lambda_2 < \lambda_3 < \ldots$ and corresponding normalized eigenfunctions $X_1, X_2, X_3, \ldots$ such that

$$LX_i = \lambda_i X_i, \; \| X_i \| = 1 \tag{5}$$
$$i = 1, 2, \ldots$$

and if "a" is any real number, then

$$X = a X_i, \; i = 1, 2, \ldots \tag{6}$$

are also solutions of (3). The norms of the solution (4) is $\| X \| = 0$ while the norm of the solution (6) is $\| X \| = a$. If we plot $\| X \|$

against λ, we find that the solution X splits into two branches at each of the eigenvalues λ_i. Therefore, the points $X = 0, \lambda = \lambda_i$ are bifurcation points of the problem.

We now consider the nonlinear eigenvalue problem

$$N(X, \lambda) = 0 \tag{7}$$

which has Equation (3) as its linearization. If we try to carry out the plot of the norms of the solutions to (7) versus λ, shown schematically in Fig. 1, we find a number of new features. For example,

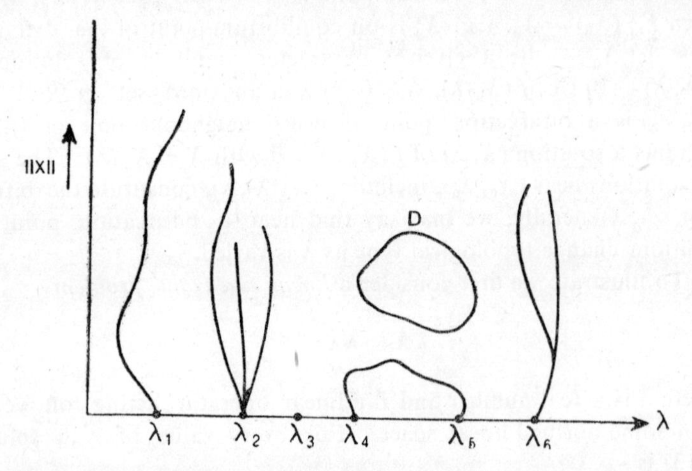

Figure 1. Branching of the solutions in nonlinear eigenvalue problem.

(1) The branches emanating from the eigenvalues of the linear problem are curved (λ_1)
(2) There may be several branches coming out of an eigenvalue (λ_2)
(3) No branch may emanate from an eigenvalue (λ_3)
(4) The branches from distinct eigenvalues of the linear problem may be connected (λ_4, λ_5)
(5) There may be branches that do not emanate from the eigenvalue of the linearized problem (piece D)
(6) There may also be a secondary bifurcation (λ_6).

3. Some Examples of Order Disorder Phenomena [7, 8]

In order to bring out the relevance of bifurcation theory for the systems of interest to us, we first try to characterize the concepts which underlie the ordering phenomena in systems such as

(a) Phase transition: ferromagnetism, superconductivity
(b) Laser operation
(c) Benard cells
(d) Chemical oscillations
(e) Neuron firings

We claim that the basic aspects of the process of spontaneous formation of organized structures in such physical, chemical or biochemical systems are analogous to the organizational aspects in living systems and in sociological pattern formations.

In the case of ferromagnetism, the system consists of a large number of interacting sub-systems which basically may be considered as electrons with spins and below a critical temperature (control parameter) $T = T_c$, the elementary magnets (spins) are lined up, giving rise to magnetization. Thus the order on the microscopic level is a cause of a new feature of the material on the microscopic level. Furthermore, to specify the organization of the subsystems, we may use the macroscopic quantity, namely, magnetization, as *order parameter*, instead of specifying the coordinates of the individual spins. A similar dramatic phase transition is observed in superconductors, which behave as a quantum system with a macroscopic wavefunction.

In a system like the laser, however, at small pump power the laser operates as a lamp—the individual atomic antennas emit light wavetrains randomly. Above a certain *threshold* pump power the individual atomic antennas oscillate in phase. Here we have an organizational situation arising out of energy flux. We note the similarity with biological processes where structural and functional order is maintained by a flux of energy and not by lowering temperature. In both cases we have thermodynamically *open systems* in *far from equilibrium condition*, working under certain natural constraints.

In the Benard cell problem again we find how the increase in a constraint (Raleigh Number) leads from a situation in which the whole of the energy is in thermal motion (random motion of individual molecules) to a much more organized state in which part of it is in the form of circular or hexagonal convective motion. This is, of course, a highly *cooperative phenomenon* from the molecular (subsystem) point of view. We may imagine that there are always small convection currents appearing as fluctuations from the average state. Below a critical value of the temperature gradient (control parameter) these fluctuations are damped and disappear. On the contrary, above some critical value certain fluctuations are amplified and give rise to a macroscopic current. A new molecular order appears that corresponds basically to a *macroscopic fluctuations stabilized by exchanges of energy with the outside world.* This may be considered as the order characterized by the occurrence of "dissipative structures".

In the system of interest to us, there is always "spontaneous noise". We recall here the role of *perturbations in stability theory.* We expect these fluctuations to play important role near "bifurcation" points. For nonlinear systems, beyond instability the statistical fluctuations increase in time and finally drive the average values to their new macroscopic state. In a number of cases such transitions to new organizational patterns may be tackled invoking Thom's theory. Thom's work is a topological bifurcation theory dealing with equations of the form

$$X = \Delta_x \, v \, (X, \lambda) \tag{8}$$

where V is a sort of potential function.

We note that in each of the systems mentioned, the equations governing "organization" are intrinsically nonlinear and have close analogies with the equations governing control circuits, e.g. in electrical engineering. From a study of these equations, we find that "modes" (such as convective or conductive transport of energy) may either compete, so that only one survives or coexist by stabilizing each other. The mode amplitude determines the degree and kind of order. We thus call them order parameters, and may point out their correspondence with order parameters in phase transition theory.

4. Basic Equations of Motion and Stability [7]

The systems under consideration, such as laser, fluid motion, etc. are described in terms of variables $q_j(t)$ or in continuous systems by $q_j(x, t)$ For example $q_j(t)$ may denote in

(i) laser: electric field strength, atomic polarization
(ii) convective instability: velocity field, fluid density
(iii) chemical reactions: densities or mole fractions of reactants

(iv) brain models: firing rates of excitatory or inhibitory neurons

All these systems have similar equations of motion which in general may be written as

$$\dot{q}_j = N_0 \, (q_j, \nabla, \lambda) + \text{damping} + \text{fluctuation} \tag{9}$$

where $N_0 \, (q_j, \nabla, \lambda)$ is a nonlinear operator. The damping and fluctuations described by fluctuating forces $F_j \, (t)$ simulate the effect of environment or internal friction, and we write

$$\dot{q}_j = N \, (q_j, \nabla, \lambda) + F_j \, (t) \tag{10}$$

In order that new structures may arise, old structures must lose stability. So one has to look into the stability aspects. We may linearize the equation using

$$q_j \rightarrow q_j^0 + \mu_j \tag{11}$$

Then the stability problem is expressed by the linearized equations

$$\dot{\mu} = \sum_m L_{jm} \, \mu_m \tag{12}$$

$$L_{jm} = L_{jm} \, (\nabla, \lambda) \tag{13}$$

are of standard form. The solutions define a set of modes $\mu_{k, l} \, (x, t)$ (**k**: wave vector; l: set of indices) whose stability is governed by parameters λ inherent in L_{jm}. Thus, when λ has acquired a certain value, we find a set of stable modes and a set of unstable modes. In the adiabatic elimination method, the stable modes may be expressed in terms of the unstable modes (time variation for stable modes considered to be negligible). The unstable modes near the critical point of instability have very long relaxation times so that they can be used as order parameters and we are left with a set of nonlinear equations for the order parameters alone. These are few, as only few degrees of freedom are highly excited.

5. The Damped Quartic Oscillator

For one degree of freedom (i.e. one order parameter) we have for a typical equation

$$\dot{q} = aq - bq^3 + F(t) \tag{14}$$

which may be interpreted as the damped motion of a fictitious particle in a potential

$$V(q) = \alpha q^2 + \beta q^4 \tag{15}$$

The fictitious particle may describe, e.g. the amplitude of laser light, or the amplitude of the velocity field of convection or a concentration wave in a chemical reaction. For example, in a laser the electric field amplitude E obeys

$$\dot{E} = a(\sigma - \sigma_t) E - b\sigma E^3 \tag{16}$$

where a, σ, etc. are known parameters and the probability

$$P(E) = N_0 \exp(- G/K_L \sigma) \tag{17}$$

where
$$G = - \tfrac{1}{2} A (\sigma - \sigma_t) E^2 + \tfrac{1}{4} B \sigma E^4 \tag{18}$$

Also in the Ginzberg-Landau theory the free energy may be written in terms of an order parameter ψ as

$$F = F_N + \alpha |\psi|^2 + \tfrac{1}{2}\beta |\psi|^4 \tag{19}$$

where F_N is the free energy of normal state, α, β are regular functions of temperature T and $|\psi|^2 \to 0$ as $T \to T_c$. Putting $\delta F/\delta\psi = 0$ yields $|\psi|^2 = -\alpha/\beta$.

Looking back at our quartic oscillators we find that for

$\beta > 0$ the system is globally stable

$\beta < 0$ leads to global instability

Furthermore, considering steady state $\dot{q} = \ddot{q} = 0$, the states of stable or unstable equilibrium are given by

$$\alpha q + \beta q^3 = 0 \tag{20}$$

For $\alpha > 0$, stable solution $q_0 = 0$, an *attractor*

$\alpha < 0$, state $q_0 = 0$ becomes unstable (*repeller*)

Considering transition from $\alpha > 0$ to $\alpha < 0$, the system passes through an instability by which the *symmetry is broken.* $\alpha = 0$, $q = 0$ is a marginal state, i.e. state of neutral stability. We may generate a sequence of potential by changing an external parameter allowing for one stable state, for one stable and one metastable and for a new stable state. Thus, the system having this type of response under external control parameters (such as magnetic field or pH value) when taken through a cyclic change of the parameter shows hysteresis. However, the bifurcation diagram (Fig. 2a) gets changed to a saddle-node type (Fig. 2b) when there is a small uncertainty ϵ in the parameter α. A model for such case is given by the bifurcations of the fixed points of Duffing's equation [9].

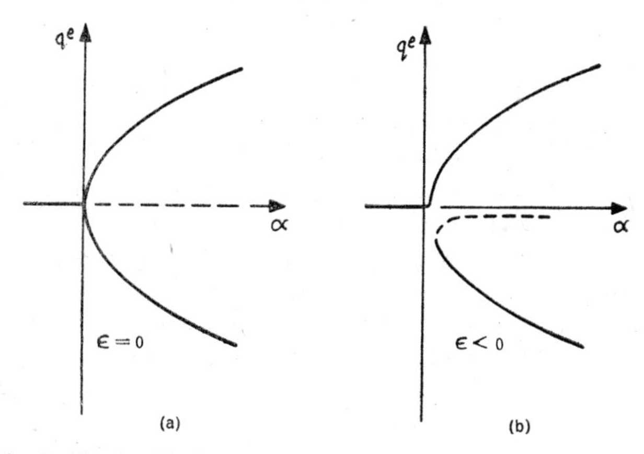

Fig. 2. The equilibrium position as a function of α (for $=0$, and $\neq 0$ but <0).

We would like to add here [12] that stability refers basically to the global properties of the phase space trajectories under perturbations as compared to the global properties of the corresponding trajectories of the unperturbed system. Thus considering again the case of one dimensional harmonic oscillator, we find that the undamped system is structurally unstable because the introduction of slight damping changes the topological properties of the trajectories.

6. Chemical Reactions [8]

The emergence of ordered behaviour in systems far from thermodynamic equilibrium has received considerable attention. A class of systems involving nonlinear chemical reactions of the auto-catalytic or cross-catalytic type including diffusion has been well studied by Nicolis, Prigogine and others. The evolution equation is

$$\frac{\partial x_i}{\partial t} = v_i\,(x_1, \ldots, x_n; \lambda) + D_i\, \nabla^2 x_i,\ i = 1, 2, \ldots n \tag{21}$$

where x_i represent concentration of chemical reactants. Under natural boundary conditions (zero flux or periodic) the solutions show a number of transition phenomena around the spatially uniform and time independent state. These transitions occur through a bifurcation mechanism involving an exchange of stability between an initially stable reference state and a new stable branch state. This is shown very clearly in reference [8, Fig. 7.11].

The most common bifurcations are those leading to

(i) multiple steady states and hysteresis without any change in spatial or temporal symmetrics,

(ii) time symmetry breaking associated with time periodic solutions of the limit cycle type,

(iii) space symmetry breaking associated with the emergence of space order.

Thus fluctuatious, which are damped when the state of the system enjoys asymptotic stability, attain macroscopic values in the vicinity of and past the bifurcation points.

The same authors [8] have discussed the following nonlinear reaction scheme:

$$A + 2X \times \underset{k_2}{\overset{k_1}{\rightleftarrows}} 3X$$

$$X \underset{k_4}{\overset{k_3}{\rightleftarrows}} B \tag{22}$$

the overall reaction being $A \rightleftarrows B$. They have shown further, how the linear stability diagram in parameter space can be interpreted as a cusp catastrophe [Fig. 8.4a].

7. Nuclear Scattering

Cusp phenomena in scattering has been treated by Wigner long a time back [10, 11]. In a number of scattering phenomena and reactions when a new channel opens up above a threshold energy Q, the phase shift may be written as

$$\delta = \delta_Q - c\,(Q - E)^{1/2} \tag{23}$$

where δ_q is the value of δ at energy $E = Q$. Since below the threshold δ has to be real, c must be real. Above threshold, δ will be complex and

$$\delta = \delta_Q - ic\,(E - Q)^{1/2} \tag{24}$$

The cross-section may be written as

$$\sigma = \frac{\pi}{k^2}\,|\,e^{2i\delta} - 1\,|^2 = \frac{4\pi}{k^2}\,|\,e^{i\delta} \sin\delta\,|^2 \tag{25}$$

where k is wave number. Then in terms of σ at $E = Q$, we get

$$\sigma = \sigma_Q\,[\,1 - 2c \cot\delta_Q\,(Q - E)^{1/2}] \qquad \text{for } E < Q$$

and

$$\sigma = \sigma_Q\,[1 + 2c\,(E - Q)^{1/2}] \qquad \text{for } E > Q$$

These expressions show that $d\sigma/dE$ is infinite at $E = Q$. One also finds that

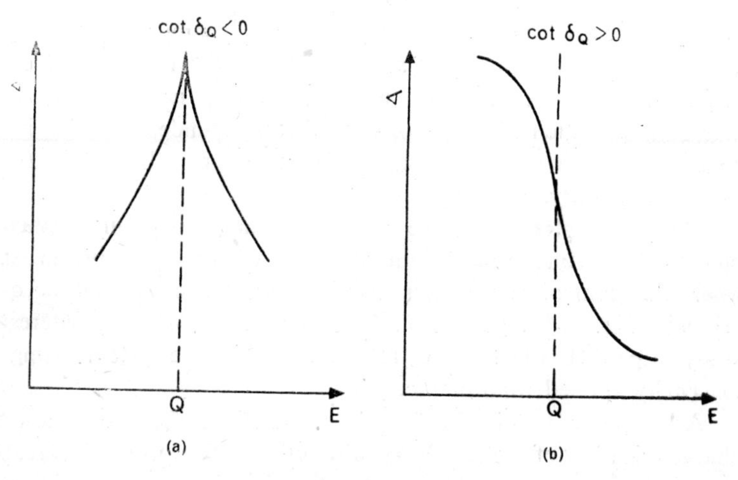

Figure 3. Plot of scattering cusp.

$$\underset{\varepsilon \to 0}{\text{lt}} \frac{\left(\frac{d\sigma}{dE}\right)_{E=Q-\varepsilon}}{\left(\frac{d\sigma}{dE}\right)_{E=Q+\varepsilon}} = \cot \delta_Q$$

which leads to cusps (see Figures 3a and b). If, however, the new channel opens up in a higher angular momentum state l, one finds that

$$\delta = \delta_Q - c\,(Q - E)^{l+1/2}, \text{ for } Q > E$$

$$\delta = \delta_Q - ic\,(E - Q)^{l+1/2}, \text{ for } Q < E$$

In this case $d\sigma/dE$ does not go to infinity but the $(l + 1)$th derivative does. Then it is difficult to observe cusp. This means that when a cusp is observed, the new channel opens up in the s-wave ($l = 0$) state. Work is in progress towards possible applications in heavy ion inelastic scattering and reactions.

8. Concluding Remarks

We have not discussed the various aspects of stability theory. Stability aspects of dynamical systems are usually studied in terms of bifurcations of potential energy functions. For fluid dynamics, the functions whose bifurcations are studied are the stream functions. For scattering one studies the bifurcations of phase.

The problem of regulation and control encountered in biology and in engineering sciences are dynamically equivalent to the study of asymtotically stable critical points of dynamical systems [12]. The characteristic property of asymptotic stability refers to an entire neighbourhood of a critical point. Within this neighbourhood the final state to which all trajectories tend is independent of the initial conditions which specify any particular trajectory. In biology this is called *equifinality*.

Many examples of bistable systems are there which jump catastrophically from one equilibrium state to another equilibrium state when the control parameters of the system are changed across a critical value. M. Agu and Y. Teramachi have done an interesting study to predict catastrophic transitions in bistable systems using an externally applied random face [13].

As opposed to the study of organization, the generation and characterization of "chaos" can also be studied using bifurcation theory [14].

We would like to conclude our discussion by referring to another very interesting application of bifurcation theory to a physical system. Constantinescu [15] has used recently bifurcation theory to study the transition from a low density nondegenerate electron gas to an "electron solid" in a superstrong magnetic field. The dynamics of the gas are described by a set of nonlinear equations and the transition is seen as the bifurcation of spatially periodic solutions from the trivial constant solution.

REFERENCES

1. Joseph B. Kellar and Stuart Autman (1969) Bifurcation Theory and Nonlinear Eigenvalue Problems, W.A. Benjamin, New York.
2. A.A. Andronov, A.A Vit and C.E. Khaikin (1966) Theory of Oscillators, Pergamon, Oxford.
3. A.A. Andronov, E.A. Leontovich, I.I. Gordon and A.G. Maier (1971) Theory of Bifurcations of Dynamic Systems on a Plane, Wiley, New York.
4. D. Sattinger (1973) Topics in Stability and Bifurcation Theory, Lecture Notes 309, Springer, Berlin.
5. D.D. Joseph (1973) Lecture Notes in Mathematics, 322 Springer, Berlin.
6. J.E. Marsden and P.J. Holmes (1978) Lecture Notes in Mathematics 648, p. 163, Springer, Berlin.
7. H. Haken (1977) Synergetics, Springer, Berlin.
8. G. Nicolis and I. Prigogine (1977) Self-Organization in Nonequilibrium Systems, Wiley, New York.
9. P. Holmes and D Rand (1976) J. Sound and Vib. 44, 237.
10. E. Wigner (1948) Phys. Rev. 73, 1002.
11. Y. Yamaguchi (1951) Prog. Theor. Phys. (Japan).
12. R. Rosen (1970) Dynamical Systems Theory in Biology, Vol. 1, Wiley, New York.
13. M. Agu and Y. Teramachi (1978) J. Appl. Phys. 49, 3645.
14. Tomoji Yamada and H. Fujisaka (1978) Prog. Theoretical Physics (Japan), Supplement 64.
15. D.H. Constantinescu (1979) Phys. Rev. Lett. 43, 1267.

Bibliography of Catastrophe theory

T. POSTON
I.N. STEWART

All the items listed in this section either make explicit reference to catastrophe theory or to very closely associated areas of mathematics, or are 'implict' applications of catastrophe theory to the sciences. The coverage of the first category is as complete as we have been able to make it, though doubtless not exhaustive; for the second category we have been more selective listing only papers especially relevant to the material discussed. (The inclusion of an item in this list does not imply that its content is without error: a small proportion of the items listed in fact contain serious mistakes. Rather than prejudice the issue, we leave it to the reader to form his own opinions.)

Abraham, R. Predictions for the future of differential equations In Symposium on Differential Equations and Dynamical Systems, Warwick 1968–69. Lecture Notes in Mathematics 206 (D.R.J. Chillingworth, ed.). Springer, Berlin and New York, 1971. pp. 163–166.

Abraham, R. Introduction to morphology. In Quatrieme Rencontre entre Mathematiques et Physique 1972, Vol, 4, fasc. 1. Mathematics Department, University of Lyons, 1972. Tome 9, pp. 38–114.

Abraham, R. and Robbin, J.W. Transversal Mappings and Flows. Benjamin, New York, 1967.

Amson, J.C. Equilibrium and catastrophic modes of urban growth. In London Papers in Regional Science 4 (E.L. Cripps, ed.). Pion, London, 1973. pp. 108–128.

Amson, J.C. Catastrophe theory: a contribution to the study of urban systems? Environ. Planning B 2, 177–221, 1975.

Anon. How catastrophe may teach us all the wrong lessons. Times Higher Educational Supplement, 5th December 1975. p.12.

Anon. The magnificent seven. Manifold 14, 6–13, 1973.

Anon. Catastrophes in action. Manifold 14, 26–31, 1973.

Anon. Bibliography. Manifold 14, 36–37, 1973.

Antonelli, P. Transporting a pure mathematician into theoretical biology. Proceedings of the Conference on Mathematics, Statistics, and the Environment Ottawa, 1974.

Arnol'd, V.I. Singularities of smooth maps, Uspehi Mat. Nauk 23, 3–44, 1968; Russian Math. Surveys 23, 1–43, 1968.

Arnol'd, V.I. On braids of algebraic functions and the cohomology of swallowtails. Uspehi Mat. Nauk 23, 247–248, 1968.

Arnol'd V.I. On matrices depending on parameters. Uspehi Mat. Nauk 26, 101–114, 1971: Russian Math. Surveys 26, 29–43, 1971.

Arnol'd, V.I. Lectures on bifurcations and versal families. Uspehi Mat. Nauk 27, 119–184, 1972; Russian Math. Surveys 27, 54–123, 1972.

Arnol'd, V.I. Integrals of rapidly oscillating functions and singularities of projections of Lagrangian manifolds, Funkcional Anal. i Prilozen 6, 61-62 1972; Functional Anal. Appl. 6, 222–224, 1972.

Arnol'd, V.I. Normal forms for functions near degenerate critical points, the Weyl groups of A_h, Dh and F_h, and Lagrangian singularities. Funkcional. Anal. i Prilozen 6, 3–25, 1972; Functional Anal. Appl. 6, 254–272, 1972,

Arnol'd, V.I. Classification of unimodal critical points of functions. Funkcional. Anal. i Prilozen. 7, 75–76, 1973; Functional Anal. Appl. 7, 230–231, 1973.

Arnol'd, V.I. Remarks on the stationary phase method and Coxeter numbers. Russian Math. Surveys, 28, 19–48, 1973.

Arnol'd, V.I. Normal forms for functions in the neighbourhood of degenerate critical points. Uspehi Mat. Nauk 29, 11–49, 1974; Russian Math. Surveys 29, 10–50, 1974.

Arnol d, V.I. Critical points of smooth functions, Proc. Internat. Congr. Math. Vancouver 1974, pp. 19–39.

Arnol'd V.I. Critical points of smooth functions and their normal forms. Uspehi Mat. Nauk 30, 3–65, 1975; Rassian Math. Surveys 30, 1–75, 1975.

Arnol'd, V.I. Classification of bimodal critical points of functions. Funkcional. Anal i. Prilozen. 9, 49–50, 1975; Functional Anal. Appl. 9, 43–44, 1975.

Arnol'd, V.I. Local normal forms of functions. Invent. Math. 35, 87–109, 1976.

Arnol'd, V.I. Wave front evolution and equivariant Morse lemma. Commun. Pure Appl. Math. 29, 557–582, 1976.

Ascher, E. and Poston, T. Catastrophe theory in scientific research. Research Futures 2, 15–18, 1976, Battelle Memorial Institute, Battelle, Ohio.

Ascher, E., Gay, D.A. and Poston, T. Equivariant bifurcation of thermodynamic potentials in crystallography. To appear.

Ashton, K. Catastrophe theory, applications in biology. Duplicated seminar notes, University of Auckland, 1976.

Atkinson, G. Catastrophe theory in geography—a new look at some old problems. Mimeographed notes, 1976.

Baas, N.A. On the models of Thom in biology and morphogenesis. Preprint, University of Virginia, 1972.

Banchoff, T.F. Polyhedral catastrophe theory 1. Maps of the line to the line. In Dynamical Systems. Proceedings of the Symposium at Salvador, Brazil, 1971 (M.M. Peixoto, ed.). Academic Press, New York and London, 1973. pp. 7–22.

Banchoff, T. and Strauss, C. A reinvestigation of the centro-surface of the ellipsoid. To appear in Francis, G.K. (ed.). Graphic Techniques in Geometry and Topology, Proc. of Special Session, Amer. Math. Soc. Evanston, Illinois, April 1977.

Battro, A.M. Morphogenese des Limnees, adaptation vitale et theorie des catastrophes. Bli de Psychologie, Hommage a Jean Piaget, Paris 1977.

Battro, A.M. Reflexions sur une psychologic ecologique experimental. Le probleme de l'echella de l'environmental. L' Annee Psychologique, 1977.

Battro, A.M. Stabilite structurelle et Psychogenese. To appear.

Battro, A.M. Le geometria de la inestabilidad y la teoria de las catastrofes, Criterio (Buenos Aires) no. 1722, 463–468, 1975.

Beer, M. Endliche Bestimmtheit und universelle Entfaltungen von Keimen mit Grup-penoperation. Diplomarbeit, University of Regensburg, 1976.

Bell, G.M. and Lavis, D.A. Thermodynamic phase changes and catastrophe theory. To appear.

Benguigui, L. and Schulman, L.S. Topological classification of phase transitions. Phys. Lett. A 45, 315–316, 1973.

Berry, M.V. The Diffraction of Light by Ultrasound. Academic Press, New York and London, 1966.

Berry, M.V. Attenuation and focussing of electromagnetic surface waves rounding gentle bends. J. Phys. A 8, 1952–1971, 1975.

Berry, M.V. Cusped rainbows and incoherence effects in the rippling-mirror model for particle scattering from surfaces. J. Phys. A 8, 566–584, 1975.

Berry, M.V. Catastrophes and semi-classical mechanics. In Rencontre de Cargese sur les Singularites et leurs Applications (F. Pham. ed.). Institut d'Etudes Scientifiques de Cargese, Publ. Math. Dept. Univ. Nice, 1975. pp. 133–136.

Berry, M.V. Waves and Thom's theorem. Adv. Phys. 25, 1–25, 1976.

Berry, M.V. Semi-classical mechanics in phase space. Phil. Trans. Roy. Soc. To appear.

Berry, M.V. Focusing and twinkling: critical exponents from catastrophes in non-Gaussian random short waves. Preprint, Univ. of Bristol, 1977.

Berry, M.V. and Hannay, J.H. Umbilic points on Gaussian random surfaces. Preprint, Univ. of Bristol, 1977.

Berry, M.V. and Mackley, M.R. The six-roll mill: unfolding an unstable persistently extensional flow. Phil. Trans. Roy. Soc., in press.

Berry, M.V. and Mount, K.E. Semiclassical approximations in wave mechanics. Rep. Prog. Phys. 35, 315–397, 1972.

Berry, M.V. and Nye, J.F. Fine structure in caustic junctions, Nature, Lond. 267, 34–36, 1977.

Bierstone, E. Local properties of smooth maps equivariant with respect to finite group actions. J. Diff Geom. 10, 523–540, 1975.

Boardman, J.M. Singularities of differentiable mappings. Publ. Math. IHES 33, 21–57, 1967.

Bochnak, J. and Kuo. T.-C. Different realizations of a non-sufficient jet. Indag. Math. 34, 24–31, 1972.

Brocker, Th. Differentierbare Abbildungen. Lecture notes, Uuiversity of Regensburg, 1973.

Brocker, Th. and Lander, L. Differentiable Germs and Catastrophes. London Mathematical Society Lecture Notes 17. Cambridge University Press, Cambridge, 1975.

Brown, B.L., Inouye, D., Williams, R. and Borrus, K. A catastrophe theory account of dichotic listening. Preprint, Dept. of Psychology, Stanford University, 1976.

Callahan, J.J. Singularities of plane maps. Amer. Math. Monthly 81, 211–240, 1974.

Callahan. J.J. Singularities of plane maps II. Sketching catastrophes. Preprint, University of Warwick, 1976, Amer. Math. Monthly, in press.

Callahan, J.J. The geometry of $E_6 = x^3 + y^4$, anorexia and the methnd of tableaus for visualizing five dimensional objects. To appear in Francis, G.K. (ed.) Graphic Techniques in Geometry and Topology, Proc. of Special Session, Amer. Math. Soc. Evanston, Illinois, April 1977.

Carpenter, G.A. Travelling wave solutions of nerve impulse equations. Thesis, University of Wisconsin, 1974.

de Carvalho, M.S.B. Liapunov functions for diffeomorphisms. Thesis, University of Warwick, 1973.

Casti, J. and Swain, H. Catastrophe theory and urban processes. Research Memorandum RM–75–14, Laxenburg, 1975.

Chapple, G. Catastrophe theory. New Zealand Listener 82, no. 1915, 16–17, 1976.

Chaudhuri, D. Rai and Jones, R.B. Unstable singularities in the σ model. J. Phys. A 9, 1349–1357, 1976.

Chazarain, J. Solutions asymptotiques et caustiques. In Rencontre de Cargese sur les Singularites et leurs Applications (F. Pham, ed.). Institut d'Etudes Scientifiques de Cargese, Publ. Math. Dept., Univ. Nice, 1975, pp. 43–78.

Chenciner, A. Travaux de Thom et Mather sur la stabilite topologique. Sem. Bourbaki no. 424, 1972–73.

Chenciner, A. Singularites des applications differentiables et catastrophes elementaires. In Rencontre de Cargese sur les Singularites et leurs Applications (F. Pham, ed.). Institut d'Etudes Scientifiques de Cargese, Publ. Math. Dept. Univ. Nice, 1975. pp. 1–5.

Chillingworth. D.R.J. Elementary catastrophe theory. Bull. Inst. Math. Appl. 11, 155–159, 1975.

Chillingworth. D.R.J. The catastrophe of a buckling beam. In Dynamical Systems—Warwick 1974. Lecture Notes in Mathematics 468 (A. Manning. ed.) Springer, Berlin and New York, 1975. pp. 86–91.

Chillingworth, D.R J. Differential Topology with a View to Applications. Research notes in mathematies 9, Pitman Publishing, London, 1976.

Chillingworth, D. R. J. (ed.) Catastrophe Theory in Infinite Dimensions. To appear.

Chillingworth, D.R.J. and Furness, P. Reversals of the earth's magnetic field. In Dynamical Systems—Warwick 1974. Lecture Notes in Mathematics 468 (A. Manning, ed.). Springer, Berlin and New York, 1975. pp. 91–98.

Chilver, H. Wider implications of catastrophe theory. Nature, Lond. 254, 381, 1975.

Chow, S.-N , Hale, J.K. and Mallet-Paret. J. Applications of generic bifurcation I. Arch. Rat. Mech. Anal. 59, 159–188, 1975.

Chow, S.-N., Hale, J.K. and Mallet-Paret, J. Applications of generic bifurcation II. Arch. Rat. Mech. Anal. 62, 209–236, 1976.

Connor, J.N.L. Multidimensional canonical integrals for the asymptotic evaluation of the *S*-matrix in semiclassical collision theory, Faraday Discuss. Chem. Soc. 55, 51–58, 1973.

Connor, J. N. L. Semiclassical theory of molecular collisions: three nearly coincident classical trajectories. Molec. Phys. 26, 1217–1531, 1973.

Connor, J.N.L. Evaluation of multidimensional canonical integrals in semiclassical collision theory. Molec. Phys. 26, 1371-1377, 1973.

Connor, J.N.L. Semiclassical theory of molecular collisions: real and complex valued classical trajectories for collinear atom Morse oscillator collisions. Molec. Phys. 28. 1569–1578, 1974.

Connor, J.N.L. Catastrophes and molecular collisions. Molec. Phys. 31, 33–55, 1976.

Cooke, J. and Zeeman, E.C. A clock and wavefront model for control of the number of repeated structures during animal morphogenesis. J. Theoretical Biology 58, 455–476, 1976.

Courant, R Soap film experiments with minimal surfaces. Amer. Math. Monthly 47, 168–174, 1940.

Croll, J. Is catastrophe theory dangerous? New Scientist, 17 June 1976, pp. 630–632.

Damon, J. Comparing topological and C^∞ stability. In Rencontre de Cargese sur les Singularites et leurs Applications (F. Pham. ed.). Institut d'Etudes Scientifiques de Cargese, Publ. Math. Dept., Univ. Nice, 1975. pp. 137–142.

Deakin. M.A.B. The formal power series approach to elementary catastrophe theory. Preprint, Monash University, 1976.

Delany, S.R. Triton. Bantam Books, New York, 1975.

Dendrinos, D.S. Mode choice, transport pricing and urban form. Mimeographed, 1975.

Dickson, D. and Thom, R. Was Newton's apple a cups or a swallowtail? Times Higher Education Supplement, 5th December 1975, p. 13.

Dobson, C.T.J. and Dodson, M.M. Simple non-linear systems and the cusp catastrophe. Preprint, York University, 1974.

Dobson, M.M. Evolution and the fold catastrophe. In Rencontre de Cargese sur les Singularites et leurs Applications (F. Pham, ed.). Institut d'Etudes Scientifiques de Careges, Publ. Math. Dept., Univ. Nice, 1975. pp. 126–127.

Dobson, M.M. Darwin's law of natural selection and Thom's theory of catastrophes. To appear in Math. Biosci.

Dodson, M.M. Quantum evolution and the fold catastrophe. To appear.

Dobson, M.M. and Hallam, A. Allopatric speciation and the fold catastrophe. To appear.

Dubosis, J.-G. and Dufour, J.-P. La theoric des catastrophes I. La machine a catastrophes. Ann. Inst. Henri Poincare 20, 113–134, 1977.

Dubois, J.-G. and Dufour, J.-P. La theorie des catastrophes II. Dynamiques gradientes a une variable d'etat. Ann. Inst. Henri Poincare 20, 135–151, 1974.

Dubois, J.-G. and Dufour, J.-P. La theorie des catastrophes V. Transformees de Legendre et thermodynamique. Preprint, Dept. de Math., Univ. de Quebec, Montreal, 1977.

Dubois, J.-G., Dufour, S.-P. and Stanek. O. La theorie des catastrophes III. Caustiques de l'optique geometrique, Ann. Inst. Henri Poincare 24, 243–260, 1976.

Dubois, J.-G., Dufour, J.-P. and Stanek, O. La theorie des catastrophes IV. Deploiements universels et leurs catastrophes. Ann. Inst. Henri Poincare 24, 261–300, 1976.

Duistermant, J.J. Osclllatory integrals, Lagrange immersions, and unfolding of singularities. Commun. Pure Appl. Math, 27, 207–281, 1974.

Eisenbud, D. and Levine H. The topological degree of a finite C^∞ map germ. In Structural Stability, the Theory of Catastrophes, and Applications in the Sciences. Lecture Notes in Mathematics 525 (P.J. Hilton, ed.). Springer, Berlin and New York, 1976. pp. 90–98.

Ekeland, I. Duality in nonconvex optimization and calculus of variations. Technical Summary Report 1675, Math. Res. Center, Univ. Wisconsin, Madison 1976.

Erber, T. and Latal, H.G. A state-area principle for (magnetic) condensation processes, Bull. Acad. Royale Belgique (Sci.) 9, 1019–1042, 1967.

Erber, T. and Sklar, A. Macroscopic irreversibility as a manifestation of micro-instability. In Modern Developments in Thermodynamics (B. Gal-Or. ed). Israel University Press, Jerusalem; Wiley, New York, 1974. pp. 281–301.

Erber, T., Latal, H.G. and Harmon, B.N. The origin of hysteresis in simple magnetic systems. Adv. Chem. Phys. 20, 71–133, 1971.

Fankhauser, H.R. Katastrophentheorie-Erganzungen, Acta Phys. Austriaca 40, 377–380, 1974.

Fankhauser, H.R. Phasenubergange als Katastrophen—Ein Beispiel. Helv. Phys. Acta 47, 486–490, 1974.

Ferguson, H.R.P. Preliminary to catastrophe theory in the behavioural sciences: how to make cusp proverbs. Proceedings, Symposium on the Behavioural Sciences, Brigham Young University 1976 (A. Bergin, ed.), BYU Press, Salt Lake City, to appear.

Ferguson, J.A. Investment decisions and sudden changes in transport. Surveyor 9, 10–11, 1976.

Field, M. Transversality in *G*-manifolds. Preprint, University of Warwick, 1976.

Fogel, J.-F., Hue, J.-L. and Thom, R. La planete de l'oncle Thom, Le Sauvage, January 1977. pp. 74–80.

Forster, W. Katastrophentheorie. Acta Phys. Austriaca 39, 201–211, 1974.

Fowler, D.H. The Riemman-Hugoniot catastrophe and van der Waals' equation. In Towards a Theoretical Biology (C.H. Waddington. ed.). Edinburgh University Press, Edinburgh, Vol. 4. 1972. pp. 1–7.

Fowler, D.H. See Thom, R. Structural Stability and Morphogenesis.

Francis, G.K. (ed.). Graphic Techniques in Geometry and Topology, Proc. of Special Session. Amer. Math. Soc. Evanston. Illinois, April 1977.

Francis, G.K. From Riemann surfaces to catastrophe machines. In Francis, G.K. (ed.). Graphic Techniques in Geometry and Topology, Proc. of Special Session, Amer. Math. Soc. Evanston, Illinois, April 1977.

Fukutome, H. Theory of the unrestricted Hartree—Fook equation and its solutions IV. Progr. Theor. Phys. 53, 1320–1336, 1975.

Furutani, N. A new approach to traffic behaviour. Preprint, University of Tokyo, 1974.

Gaffney, T. Properties of finitely determined map germs. Thesis, Brandeis Univ., June 1975.

Gaffney, T. On the order of determination of a finitely determined germ. Invent. Math. 37, 83–92, 1976.

Gibson, C.G. Singular points of smooth mappings: a geometric introduction. In preparation.

Gibson, C.G., Wirthmuller, K., du Plessis, A.A. and Looijenga, E. Topological Stability of Smooth Mappings. Lecture Notes in Mathematics 552. Springer, Berlin and New York, 1977.

Gilmore, R. Structural stability of the phase transition in Dicke-like models. J. Math. Phys. A 18, 17–22, 1977.

Godwin, A.N. Elementary catastrophes. Thesis, University of Warwick, 1971.

Godwin, A.N. Three dimensional pictures for Thom's parabolic umbilic. Publ. Math. IHES 40, 117–138, 1971.

Godwin, A.N. Methods for Maxwell sets of cuspoid catastrophes. Preprint, Lanchester Polytechnic, Rugby, 1974.

Godwin, A.N. Topological bifurcation for the double cusp polynomial. Proc. Cambridge Philos. Soc. 77, 293–312, 1975.

Golubitsky, M. Contact equivalence for Lagrangian submanifolds. In Dynamical Systems—Warwick 1974. Lecture Notes in Mathematics 468 (A. Manning, ed.). Springer, Berlin and New York, 1975. pp. 71–73.

Golubitsky, M. Regularity and stability of shock waves for a single conservation law. In Rencontre de Cargese sur les Singularitiès et leurs Applications (F. Pham, ed.), Institut d'Etudes Scientifiques de Cargese, Publ. Math. Dept., Univ. Nice, 1975, pp. 84–88.

Golubitsky, M. An introduction to catastrophe theory and its applications. Lecture notes, Queens College, New York, 1976.

Golubitsky, M. and Guillemin, V. Stable Mappings and Their Singularities. Graduate Texts in Mathematics 14. Springer, Berlin and New York, 1973.

Goodwin, B. Review of Thom, R. Stabilite Structurelle et Morphogenese. Nature, Lond. 242, 207–208, 1973.

Graham, R. Phase-transition-like phenomena in lasers and nonlinear optics. In Synergetics (H. Haken, ed.). Teubner, Stuttgart, 1973, pp. 71–86.

Gray, A. A proof of the polynomial division theorems via smoothing operators. Preprint, La Trobe University, 1976.

Gromoll, D. and Meyer, W. On differentiable functions with isolated critical points. Topology 8, 361–370, 1969.

Grossman, S. Fluctuations near phase transitions in restricted geometries. In Synergetics (H. Haken, ed.). Teubner, Stuttgart, 1973, pp. 54–70.

Guckenheimer, J. Caustics. In Proceedings of the UNESCO Summer School, Trieste 1972. International Atomic Energy Authority. Vienna, pp 281–289.

Guckenheimer, J. Review of Thom, R. Stabilite Structurelle et Morphogenese. Bull. Amer. Match. Soc. 79, 878–890, 1973.

Guckenheimer, J. Bifurcation and catastrophe. In Dynamical Systems. Proceedings of the Symposium at Salvador, Brazil, 1971 (M.M. Peixoto, ed.). Academic Press, New York and London, 1973. pp. 95–110.

Guckenheimer, J. Catastrophes and partial differential equations. Ann. Inst. Fourier 23, 31–59, 1973.

Guckenheimer, J. Solving a single conservation law. In Dynamical Systems—Warwick 1974. Lecture Notes in Mathematics 468 (A. Manning, ed.). Springer, Berlin and New York, 1975. pp. 108–134.

Guckenheimer, J. Caustics and non-degenerate Hamiltonians. Topology 13, 127–133, 1974.

Guckenheimer, J. Constant velocity waves in oscillating chemical reactions. In Structural Stability, the Theory of Catastrophes, and Applications in the Sciences. Lecture Notes in Mathematics 525 (P.J. Hilton. ed.). Springer, Berlin and New York, 1976. pp. 99–103.

Guckenheimer, J. Shocks and rarefactions in two space dimensions. To appear in Arch. Rat. Mech. Anal.

Guckenheimer, J. Isochrons and phaseless sets. To appear in J Math. Biol.

Guckenheimer, J. On the bifurcation of maps of the interval. To appear.

Gusein-Zade, S.M. Intersection matrices of some singularities of functions of two variables. Funkcional. Anal. i Prilozen. 8, 15, 1974; Functional Anal. Appl. 8, 10 13, 1974.

Gnsein-Zade, S.M. Dynkin diagrams for singularities of functions of two variables. Funkcional. Anal. i Prilozen. 8, 23–30, 1974; Functional Anal. 8, 295–300, 1974.

Guttinger, W. Catastrophe geometry in physics and biology. Physics and Mathematics of the Nervous System, Lecture Notes in Biomathematics 4 Springer, Berlin and New York, 1974, pp. 2–30.

Hahn, H. Geometrical aspects of the pseudo steady state hypothesis in enzyme reactions. Physics and Mathematics of the Nervous System, Lecture Notes in Biomathematics 4 Springer, Berlin and New York, 1974. pp. 528–545.

Haken, H. (ed.), Synergetics. Taubner, Stuttgart, 1973.

Harrison, P.J. and Zeeman, E.C. Applications of catastrophe theory to macroeconomics. Symposium on Applied Global Analysis, Utrecht, 1973.

Heatley, B. Local stability properties equivalent to catastrophe theory. Thesis, University of Warwick, 1974.

Hilbert, D. Uber die Singularitaten der Diskriminantenflache. Math. Ann. 30 (1887) 437–441; Gesammelte Abhandlungen, Vol. 2, 117–120. Springer, Berlin, 1933.

Hilton, P.J. (ed.) Structural Stability, the Theory of Catastrophes, and Applications in the Sciences. Lecture Notes in Mathematics 525. Springer, Berlin and New York, 1976.

Hilton, P.J. Structural stability, catastrophe theory, and their applications to the sciences and engineering. Research Futures 1. Battelle Memorial Institute, Ohio, 1976.

Hilton. P.J. Unfolding of singularities. Colloquium on functional analysis, Campinas, Brazil, 1974.

Holford, R.I. Modifications to ray theory near cusped caustics. Preprint, Bell Telephonic Laboratories, 1972.

Holmes, P.J. and Rand, D.A. The bifurcations of Duffing's equation: an application of catastrophe theory, J. Sound Vib. 44, 237–253, 1976.

Hughes, A. An application of catastrophe theory, Math, Gazette 61, 1–20, 1977.

Inoue, M. Catastrophes and fluctuations of polarization in anisotropic dielectrics. J. Chem. Phys. 68, 3351–3354, 1976.

Inouye, D. The dynamic microstructure of evaluative processes: structural stability models of judgement and intentional action. To appear.

Isnard, C.A. and Zeeman, E.C. Some models from catastrophe theory in the social sciences. In Use of Models in the Social Sciences (L. Collins, ed.). Tavistock, London, 1976. p. 44–100.

James, I.M. Singularities and catastrophes; a sketch. Duplicated notes, Summer Research Institute, Australian Mathematical Society, Monash University, 1974.

Jänich, K. Caustics and catastrophes. In Dynamical Systems—Wawick 1974. Lecture Notes in Mathematics 468 (A. Manning, ed.). Springer. Berlin and New York, 1975. pp. 100–101.

Jänich, K. Caustics and catastrophes. Math. Ann. 209, 161–180, 1974.

Källman C.G. Lee-Wick states as an example in Thom's catastrophe theory. Phys. Lett. A 56, 70, 1976.

Kilmister, C.W. The concept of catastrophe (review of Thom, R. Stabilite Structurelle et Morphogenese). Times Higher Educational Supplement, 30th November 1973. p. 15.

Kilmister, C.W. Population in cities. Math. Gazette 60, 11–24, 1976.

King, H.C. Real analytic germs and their varieties at isolated singularities. Invent. Math. 37, 193–230, 1976.

Klahr, D. and Wallace, J.G. Cognitive Development—an Information-Processing View. Lawrence Erlbaum Associates, Hillsdale, N.J., 1976; Wiley, New York and London 1976.

Komorowski, J. On Thom's idea concerning Guggenhim's one-third law in phase transitions. Preprint, University of Warwick, 1977.

Kozak, J.J. and Benham, C.J. Denaturation; an example of a catastrophe. Proc. Nat. Acad. Sci. U.S.A. 71, 1977–1981, 1974.

Kuiper, N.H. C^1-equivalence of functions near isolated critical points. Symposium on infinite-dimensional topology (R.D. Anderson, ed.). Annals of Mathematical Studies 69, Princeton University, 1972.

Kuo, T.-C. On C^0-sufficiency of jets. Topology 81, 167–171, 1969.

Kuo, T.-C A complete determination of C^0-sufficiency in J^r (2, 1). Invent. Math. 8, 225–235, 1969.

Kuo, T.-C. The jet space J^r (n, 1). In Proceedings of the Liverpool Singularities Symposium. Lecture Notes in Mathematics 192 (C.T.C. Wall, ed). Springer, Berlin and New York, 1971. pp. 169–177.

Kuo, T.-C. Characterizations of ν-sufficiency of jets. Topology 11, 115–131, 1972.

Lacher, R.C., McArthur, R. and Buzyna, G. Catastrophic changes in circulation flow patterns. Preprint, Florida State University, 1977.

Lassalle, M.G. Une demonstration du theoreme de division pour les functions differentiable. Topology 12, 41-62, 1973.

Lassale, M.G. Deploiement universal d'une application de codimension finie. Ann. Scient Ecole Norm. Super. 7, 219–234, 1974.

Latour, F. Stabilite des camps d'applications differentiables; generalisation d'un theoreme de Mather. C.R. Acad. Sci. Paris 268, 1331, 1969.

Lax, P.D. Asymptotic solutions of oscillatory initial value problems. Duke Math. J. 24, 627–646, 1957.

Levine, H.I. Singularitites of differentiable mappings. Notes of lectures by R. Thom, Bonn, 1959; also in Proceedings of the Liverpool Singularities Symposium. Lecture Notes in Mathematics 192 (C.T.C. Wall, ed.). Springer, Berlin and York, 1971. pp. 1–89

Ljasko, O.V. Decomposition spaces of singularities of functions. Funkcional. Anal. i Prilozen. 10, 40–56, 1976.

Looijenga, E Structural stability of families of C^∞-functions and the canonical stratification of C^∞ (N). Preprint, IHES 1974.

Looijenga, E. On the semi-universal deformations of Arnol'd's unimodular singularities. Preprint, University of Liverpool, 1975.

Looijenga, E. On the semi-universal deformation of a simple-elliptic singularity part I. Unimodularity. Topology 16, 257–262, 1977.

Looijenga, E. On the semi-universal deformation of a simple-elliptic singularity Part II. Geometry of the discriminant locus. Preprint, University of Nijmegen, 1976.

Lu, Y.-C. Sufficiency of jets in J' (2, 1) via decomposition. Invent. Math. 10, 119–127, 1970.

Lu, Y -C. Singularity Theory and an Introduction to Catastrophe Theory. Springer, Berlin and New York, 1976.

Lu, Y.-C. and Chang, S.H. On C^0-sufficiency of complex jets. Can. J. Math. 25, 874–880, 1973.

Ludwig, D. Uniform asymptotic expansion at a caustic. Commun. Pure Appl. Math. 19, 215–250, 1966.

Magnus R.J. On universal unfoldings of certain real functions on a Banach space. Mathematics Report 100, Battelle, Geneva, 1976. Math. Proc. Cambridge Philos. Soc. 81, 91–95, 1977.

Magnus, R.J. Determinacy in a class of germs on a reflexive Banach space. Mathematics Report 103, Battelle, Geneva, 1976. To appear in Math. Proc. Cambridge Philos. Soc.

Magnus, R.J. On the orbits of a Lie group action. Mathematics Report 105, Battelle, Geneva, 1976.

Magnus, R.J. Universal unfoldings in Banach spaces: reduction and stability. Mathematics Report 107, Battelle, Geneva, 1977. To appear in Chillingworth, D.R.J. (ed.) Catastrophe Theory in Infinite Dimensions.

Magnus, R. and Poston, T. On the full unfolding of the von Karman equation at a double eigenvalue. Mathematics Report 109, Battelle, Geneva, 1977. To appeear in Chillingworth, D.R.J. (ed.) Catastrophe Theory in Infinite Dimensions.

Magnus. R. and Poston, T.A. Strictly infinite dimensional 'fold catastrophe'. Mathematics Report 110, Battelle, Geneva, 1977.

Malgrange, B. The preparation theorem for differentiable functions. In Differential Analysis, Bombay Colloquium. Oxford University Press, Oxford and New York, 1964. pp. 203–208.

Malgrange, B. Ideals of Differentiable Functions. Oxford University Press, Oxford and New York, 1966.

Manning, A. (ed.) Dynamical Systems—Warwick 1974. Lecture Notes in Mathematics 468. Springer, Berlin and New York, 1975.

Markus, L. Lectures in differentiable Dynamics. CBMS Regional Conference Series in Mathematics 3. American Mathematical Society, Providence, R.I., 1971.

Markus, L. Dynamical systems—five years after. In Dynamical Systems—Warwick 1974. Lecture Notes in Mathematics 468 (A. Manning, ed.). Springer, Berlin and New York, 1975. pp. 354–365.

Marsden, J.E. Applications of Global Analysis in Mathematical Physics. Lecture Notes 2. Publish or Perish, Boston, 1974.

Marsden, J.E. and McCracken, M. The Hopf Bifurcation and its Applications. Applied Mathematics Series 19. Springer, Berlin and New York, 1976.

Martinet, J. Singularities des Fonctions et Applications Differentiables. Lecture Notes. PUC, Rio de Janeiro, 1974.

Martinet, J. Deformations verselles des fonctions numeriques. Catastrophes elementaires de R. Thom. In Rencontre de Cargese sur les Singularities et leurs Applications (F. Pham. ed.). Institut d'Etudes Scientifiques de Cargese, Publ. Math. Dept., Univ. Nice, 1975. pp. 6–19.

Martinet, J. Deploiements versels des applications differentiables et classification des applications stables. In Burlet, O. and Ronga, F. (eds.). Singularities d'Applications Differentiables, Plans-sur-Bex 1975, Lecture Notes in Mathematics 535, Springer, Berlin and New York, 1976. pp. 1–44.

Martinet, J. Deploiements stables des germes de type fini, et determination finie des applications differentiables. Preprint, Math. Dept. Univ. Strasbourg, 1976.

Mather, J. Stability of C^∞-mappings I. The division theorem. Ann. Math. 87, 89–104, 1968.

Mather, J. Stability of C^∞-mappings II. Infinitesimal stability implies stability. Ann. Math. 89, 254–291, 1969.

Mather, J. Stability of C^∞-mappings III. Finitely determined map germs. Publ. Math. IHES 35, 127–156, 1968.

Mather, J. Stability of C^∞-mappings IV. Classification of stable germs by *R*-algebras. Publ. Math. IHES 37, 223–248, 1969.

Mather J. Stability of C^∞-mappings V. Transversality. Adv. Math. 4, 301–336, 1970.

Mather, J. Stability of C^∞-mappings VI. The nice dimensions. In Proceedings of the Liverpool Singularities Symposium. Lecture Notes in Mathematics 192 (C.T.C. Wall, ed.). Springer, Berlin and New York, 1971. pp. 207–253.

Mather, J. Right equivalence. Preprint, University of Warwick, 1969.

Mather, J. Notes on topological stability. Preprint, Harvard University, 1970.

Mather, J. On Nirenberg's proof of Malgrange's preparation theorem. In Proceeding of the Liverpool Singularities Symposium. Lecture Notes in Mathematics 192 (C.T.C. Wall. ed.). Springer, Berlin and New York, 1971. pp. 116–120.

Mather, J. Stratifications and mappings. In Dynamical Systems. Proceedings of the Symposium at Salvador, Brazil, 1971 (M.M. Peixoto, ed.). Academic Press, London and New York, 1973. pp. 195–232.

Mather, J. How to stratify mappings and jet spaces. In Burlet, O. and Ronga, F. (eds.). Singularities d'Applications Differentiables, Plans-sur-Bex 1975, Lecture Notes in Mathematics 535, Springer, Berlin and New York, 1976. pp. 128-176.

Max, N. New films of the butterfly catastrophe, and critical points of functions from the plane. To appear in Francis, G.K. (ed.). Graphic Techniques in Geometry Topology, Proc. of Special Session, Amer. Math. Soc. Evanston, Illinois, April 1977.

Mees, A.I. The revival of cities in medieval Europe: an application of catastrophe theory. Regional Sci. Urban Econ. 5, 403-425, 1975.

Michor, P. Classification of elementry catastrophes of codimension $\leqslant 6$. Universität Linz Institutsbericht 51, 1976.

Michor, P. The preparation theorem on Banach spaces. To appear in Chillingworth, D. R. J. (ed.). Catastrophe Theory in Infinite Dimensions.

Milnor, J. Morse Theory. Annals of Mathematical Studies 51, Princeton University, 1963.

Mitchison, G. Topological models in biology: an art or a science? MRC Molecular Biology Unit, Cambridge, 1973.

Monson, S. R. An APL implementation of Kushnirenko's Theorem to find the Milnor number of a polynomial. Preprint, Math. Dept, Brigham Young University, 1977.

Morin, B. Calcul Jacobien, Ann. Sci., Ecole Norm Super. 8, 1–98, 1975.

Morse, M. The critical points of a function of n variables. Trans. Amer. Math. Soc. 33, 72–91, 1931.

Nicolis, G. and Auchmuty, J.F.G. Dissipative structures, catastrophes, and pattern formation: a bifurcation analysis. Proc. Nat. Acad. Sci. U.S.A. 71, 2748–2751, 1974.

Nirenberg, L. A proof of the Malgrange preparation theorem. In Proceedings of the Liverpool Singularities Symposium. Lecture Notes in Mathematics 192 (C.T.C. Wall, ed.). Springer, Berlin and New York, 1971. pp. 97–105.

Noguchi, H. and Zeeman E.C. Applied Catastrophe Theory. Bluebacks, Kodansha, Tokyo, 1974. (In Japanese.)

Olenick, R. and Erber, T. A λ transition of two magnetic dipoles. Amer. J. Phys. 42, 338–339, 1974.

Palamodov, V. P. The multiplicity of a holomorphic mapping. Funkcional. Anal. i Prilozen. 1, 54–65, 1967; Functional Anal. Appl. 1, 218–226, 1967.

Panati, C. Catastrophe theory. Newsweek, 19th January 1976. pp. 46–47.

Pattee, H.H. Discrete and continuous processes in computers and brains. Physics and Mathematics of the Nervous System, Lecture Notes in Biomathematics 4, Springer, Berlin 1974, pp. 128–149.

Pearcey, T. The structure of an electromagnetic field in the neighbourhood of a cusp of a caustic. Philos. Mag. 37, 311–317, 1946.

Peixoto, M.M. (ed.) Dynamical systems. Proceedings of the Symposium at Salvador, Brazil, 1971. Academic Press, New York and London, 1973.

Pham, F. Introduction a l' Etude Topologique des Singularities de Landau. Gauthier-Villars, Paris, 1967.

Pham, F. Remarque sur l'equisingularite universelle. Preprint, University of Nice, 1970.

Pham, F. Classification des singularites. CNRS preprint Vol. 13, Strasbourg, 1971.

Pham, F. (ed.) Rencontre de Cargese sur les Singularites et leurs Applications. Institut d'Etudes Scientifiques de Cargese, Publ. Math. Dept. Univ. of Nice, Sept. 1975.

Pham, F. Caustics and microfunctions. RCP-25 23, 91–104, 1976. IRMA, CNRS Strasbourg.

Pitt, D. H. and Poston, T. Generic buoyancy and metacentric loci. To appear.

Pitt, D. H. and Poston. T. Determinacy and unfoldings in the presence of a boundary. To appear.

Poenaru, V. Analyse Differentielle. Lecture Notes in Mathematics 371. Springer, Berlin and New York, 1974.

Poenaru, V. The Maslov index for Lagrangian manifolds. In Dynamical Systems—Warwick 1974. Lecture Notes in Mathematies 468 (A. Manning. ed.). Springer, Berlin and New York, 1975. pp. 70–71.

Poenaru, V. Zakalyukin's proof of the (uni)versal unfolding theorem. In Dynamical Systems–Warwick 1974. Lecture Notes in Mathematics 468 (A Manning, ed.). Springer, Berlin and New York, 1975 pp. 85–86.

Poenaru, V. Theorie des invariantes C^∞: stabilite structurelle equivariante I and II Preprints, Orsay, 1975.

Poenaru, V. Singularites C^∞ en Presence de Symetrie. Lecture Notes in Mathematics 510. Springer, Berlin and New York, 1976.

Porteous, I.R. Geometric differentiation—a Thomist view of differential geometry. In Proceedings of the Liverpool Singularities Symposium II. Lecture Notes in Mathematics 209 (C.T.C. Wall, ed.). Springer, Berlin and New York, 1971. pp. 121–127.

Porteous, I. R. The normal singularities of a submanifold. J. Diff. Geom. 5, 543–564, 1971.

Porteous, I. R. Nobel prizes for catastrophes. Manifold 15, 34–36, 1974.

Poston, T. Do-it-yourself catastrophe machine. Manifold 14, 40, 1973.

Poston, T. Various catastrophe machines. In Structural Stability, the Theory of Catastrophes, and Applications in the Sciences. Lecture Notes in Mathematics 525 (P. J. Hilton, ed.). Springer, Berlin and New York, 1976. pp. 111–126.

Poston, T. The computational rules of catastrophe theory. In Overlapping Tendencies in Operations Research, Systems Theory, and Cybernetics (E. Billeter, M. Cuenod and S. Klaczko, eds.). Birkhauser, 1976. pp. 461–469.

Poston, T. The elements of catastrophe theory *or* the honing of Occam's razor. To appear in Cooke, K.L. and Renfrew, A. C. (eds.). Transfornnations: Mathematical Approaches to Culture Change. Academic Press, New York, 1978.

Poston, T. and Stewart, I.N. Taylor Expansions and Catastrophes. Research Notes in Mathematics 7. Pitman Publishing, London, 1976.

Poston, T. and Stewart, I. N. The geometry of binary quartic forms II. Foliation by crossratio. Preprint, University of Warwick, 1977.

Poston, T. and Wilson, A. G. Facility size vs. distance travelled: urban services and the fold catastrophe. Environ. Planning A 9 ,681-686, 1977.

Poston, T. and Woodcock, A. E. R. On Zeeman's catastrophe machine. Proc. Cambridge Philos, Soc. 74 217–226, 1973.

Poston. T., Stewart. I.N. and Woodcock, A. E. R. The geometry of the higher catastrophes. In preparation.

Put, M. van der. Some properties of the ring of germs of C^∞ functions. Compositio Math. 34, 1977.

Rand, D. Arnol'd's classification of simple singularities of smooth functions. Duplicated notes. Math. Inst. Univ. Warwick, April 1977.

Renfrew, A. C. and Poston, T. Discontinuities in the endogenous change of settlement pattern. To appear in Cooke, K. L. and Renfrew, A. C. (eds.). Transformations: Mathematical Approaches to Culture Change. Academic Press, New York, 1978

Rockwood, A. The canonical strip. Preprint, Math. Dept. Brigham Young Univ. Provo, Utah, 1976.

Rockwood, A, and Burton, R. An inexpensive technique for displaying algebraically defined surfaces. To appear in Francis, G. K. (ed.). Graphic Techniques in Geometry and Topology, Proc. of Special Session, Amer. Math. Soc. Evanston, Illinois, April 1977.

Ronga, F. Stabilite locale des applications equivariantes. Preprint, Mathematics Department, University of Geneva, 1976.

Ronga, F. and Cerveau, D. Applications topologiquement stables, Seminaire de Topologie, Dijon 1975.

Rossler, O.E. Adequate locomotion strategies for an abstract organism in an abstract environment—a relational approach to brain function. Physics and Mathematics of the Nervous System, Lecture Notes in Biomathematics 4. Springer, Berlin and New York, 1974. pp. 342–369.

Rossler, O. E. A synthetic approach to enzyme kinetics. Physics and Mathematics of the Nervous System, Lecture Notes in Biomathematics 4. Springer, Berlin and New York, pp. 546–582.

Rozestraten, R. J. A., Battro, A. M. and Dos Santos Andrade, A. A visual catastrophe: the reversal of the Oppel-Kundt illusion in the open field. Abstract guide of the XXIst International Congress of Psychology, Paris, July 1976, p. 316.

Ruelle, D. and Takens, F. On the nature of turbulence. Commum. Math. Phys. 20, 167–192, 1971.

Schulman, L. S. Phase transitions as catastrophes. In Symposium on Differential Equations and Dynamical Systems, Warwick 1968–69. Lecture Notes in Mathematics 206 (D. R. J. Chillingworth, ed.). Springer, Berlin and New York, 1971. pp. 98–100.

Schulman, L. S. Tricritical points and type three phase transitions. Phys. Rev. Ser. B 7, 1960–1967, 1973.

Schulman, L.S. Stable generation of simple forms. J. Theor. Biol. 57, 453–468, 1976.

Schulman, L. S. and Revzen, M. Phase transitions as catastrophes. Collect. Phenom. 1, 43–47, 1972.

Schwarz, G. Smooth functions invariant under the action of a compact Lie group. Topology 14, 63–68, 1975.

Sergeraert, F. La stratification naturelle de C^{∞} (M). Thesis, Orsay. 1971.

Setterstrom, R. D. The double Zeeman catastrophe machine. Preprint, Math. Dept Brigham Young Univ. Provo, Utah, 1976.

Sewell, M. J. On the connexion between stability and the shape of the equilibrium surface. J. Mech. Phys. Solids 14, 203–230, 1966.

Sewell, M. J. Kitchen catastrophe. Math. Gazette 59, 246–249, 1975.

Sewell, M. J. Some mechanical examples of catastrophe theory. Bull. Inst. Math. Applic. 12, 163–172, 1976.

Sewell, M. J. Elementary catastrophe theory. In Proceedings of the International Conference on Problem Analysis in Science and Engineering, Waterloo University 1975, to appear.

Sewell, M. J. Review of Thom, R. Stabilite Structurelle et Morphogenese. Math. Gazette, in press.

Sewell, M. J. On Legendre transformations and elementary catastrophes. Technical Summary Report, 1707, Math. Res. Center, University of Wisconsin, Madison, 1976.

Sewell, M. J. Elastic and plastic bifurcation theory. Preprint, University of Reading, 1976.

Sewell, M. J. Some global equilibrium surfaces. Technical Summary Report, 1714, Math. Res. Center, Univ. of Wisconsin, Madison, 1977.

Sewell, M. J. A survey of plastic buckling. In Study No. 6 (Stability), Solid Mechanics Division, University of Waterloo, Ontario, Canada, 1972. Chap. 5.

Sewell, M. J. Degenerate duality, catastrophes and saddle functionals. Preprint, Univ. of Reading, 1977.

Siersma, D. Singularilies of C^{∞} functions of right-codimension smaller or equal than eight. Indag. Math. 25, 31–37, 1973.

Siersma, D. Classification and deformation of singularities. Thesis, Amsterdam, 1974.

Smale, S. On gradient dynamical systems. Ann. Math. 74, 199–206, 1961.

Smale, S. Differentiable dynamical systems. Bull. Amer. Math. Soc. 73, 747 817. 1967.

Smale, S. Global analysis and economics I. Pareto optimum and a generalisation of Morse theory. In Dynamical Systems. Proceedings of the Symposium at Salvador, Brazil, 1971 (M. M. Peixoto, ed.). Academic Press, New York and London, 1973. pp. 531–544.

Smale, S. Global analysis and economics II. Extension of a theorem of Debreu. J. Math. Econ. 1, 1–14, 1974.

Smale, S. Pareto optima and price equilibria. To appear.

Smith, T. R. Continuous and discontinuous response to smoothly decreasing effective distance: an analysis with special reference to 'overbanking' in the 1920's. To appear in Environ. Planning A 9.

Stanley, H. E. Introduction to Phase Transitions and Critical Phenomena. Oxford University Press, London and New York, 1971.

Starobin, L. Our changing evolution: strategies for 1980. General Systems 21, 3–46, 1976.

Stefan, P. A remark on right k-determinacy. Preprint, Bangor University, 1974.

Stewart, I. N. Concepts of Modern Mathematics. Penguin, Harmondsworth, Middlesex, 1974.

Stewart, I. N. The seven elementary catastrophes. New Scientist 68, 447–454, 1975.

Stewart I. N. The geometry of binary quartic forms part I. Preprint, University of Warwick, 1976.

Stewart, I. N. Catastrophe theory. Math. Chronicle, 5, 140–165, 1977.

Stewart, I. N. Catastrophe Theory. In Encyclopaedia Britannica, Special Supplement to the Yearbook 1977. To appear.

Sussmann, H. J. Catastrophe theory. Synthese 31, 229–270, 1975.

Sussmann, H. J. Catastrophe theory—a preliminary critical study. To appear in Biennial Meeting of the Philosophy of Science Association, Chicago, October 1976

Sussmann, H. J. and Zahler, R. S. Catastrophe theory as applied to the social and biological sciences: a critique. To appear in Synthese.

Takens, F A note on sufficiency of jets. Invent. Math. 13, 225–231, 1971.

Takens, F. Singularities of functions and vector fields. Nieuw. Arch. Wisk. 20, 107–130, 1972.

Takens, F. Introduction to global analysis. Mathematics Institute, Utrecht University, 1973.

Takens, F. Constrained differential equations. In Dynamical Systems—Warwick 1974. Lecture Notes in Mathematics 468 (A. Manning, ed.) Springer, Berlin and New York, 1975. pp. 80–82.

Takens, F. Constrained equations: a study of implicit differential equations and their discontinuous solutions. In Structural Stability, the Theory of Catastrophes, and the Application in the Sciences. Lecture Notes in Mathematics 525 (P. J. Hilton, ed.). Springer, Berlin and New York, 1976, pp. 143–234.

Takens, F. Implicit differential equations: some open problems. In Singularites d'Applications Differentiables, Plans-sur-Bex, 1975. (Burlet, O. and Ronga, F. eds.), Lecture Notes in Mathematics 535, Springer, Berlin and New York, 1976. pp. 237–253.

Tall, D. O. A long term learning schema for calculus and analysis. Math. Educ. Teachers 2, No. 2, 1975.

Tall, D. O. Conflicts and catastrophes in the learning of mathematics. Math. Educ. Teachers 3, No. 2, 1976.

Teissier, B. Sur la version catastrophique de la regle des phases de Gibbs et l'invariant δ des singularites d'hypersurfaces. In Rencontre de Cargese sur les Singularites et leurs Applications (F. Pham, ed.). Institut d'Etudes Scientifiques de Cargese, Publ. Math. Dept., Univ. Nice, 1975, pp. 105–113.

Thom, R Une lemme sur les applications differentiables. Bol. Soc. Mat. Mexicana 1, 59–71, 1956.

Thom, R Les singularites des applications differentiables. Ann. Inst. Fourier 6, 43–87, 1956.

Thom, R. La stabilite topologique des applications polynomiales. L'Enseignement Math. 8, 24–33, 1962.

Thom, R. Sur la theoric des enveloppes, J. Math. Pures Appl. 41, 177–192, 1962.

Thom, R. Local properties of differentiable mappings In Differential Analysis, Bombay Colloquium. Oxford University Press, Oxford and New York, 1964.

Thom, R. L'equivalence d'une fonction differentiable et d'un polynome. Topology 3, 297–307, 1965.

Thom, R. On some ideals of differentiable functions J. Math. Soc. Japan 19, 255–259, 1967.

Thom, R. Topologic et signtficaution. L'Age Sci. 4, 219–242, 1968.

Thom, R. Comments on C.H. Waddington: the basic ideas of biology. In Towards a Theoretical Biology (C. H. Waddington, ed.). Edinburgh University Press, Edinburgh, Vol 1, 1968. pp. 32–41

Thom, R. Une theorie dynamique de la morphogenese. In Towards a Theoretical Biology (C. H. Waddington, ed.). Edinburgh University Press, Edinburgh, Vol. 1, 1968. pp. 152–179.

Thom, R. A mathematical approach to morphogenesis: archetypal morphologies. In Heterospecific Genome Interaction. Wistar Institute Symposium Monograph 9. Wistar Institute Press, Tel Aviv, 1969.

Thom, R. Topological models in biology. Topology 8, 313–335, 1969; also in Towards a Theoretical Biology (C. H. Waddington, ed.). Edinburgh University Press, Edinburgh, Vol. 3, 1970. pp. 89–116.

Thom, R. Ensembles et morphismes stratifies. Bull. Amer. Math. Soc. 75, 240–284, 1969.

Thom, R. Sur les varietes d'ordre fini. In Global Analysis (Papers in Honour of K. Kodaira). Tokyo, 1969. pp. 397–401.

Thom, R. The bifurcation subset of a space of maps. In Manifolds, Amsterdam 1970 (N. H. Kuiper, ed.). Lecture Notes in Mathematics 197. Springer, Berlin and New York, 1971. pp. 202–208.

Thom, R Topologie et linguistique. In Essays on Topology and Related Topics (Dedicated to G. de Rham) (A. Haefliger and R. Narasimhan. eds). Springer, Berlin and New York, 1970. pp. 226–248.

Thom, R. Les symmetries brisees en physique macroscopique et la mecanique quantique. CNRS, RCP-25 10, 1970.

Thom, R. Singularities of differentiable mappings. See Levine, H. I.

Thom, R. Stratified sets and morphisms: local models. In Proceedings of the Liverpool Singularities Symposium. Lecture Notes in Mathematics 192 (C. T. C. Wall, ed.). Springer, Berlin and New York, 1971. pp. 153–164.

Thom, R. Sur le cut-locus d'un variete plongee. J. Diff. Geom. 6, 577–586, 1972.

Thom, R Structuralism and biology. In Towards a Theoretical Biology (C. H. Waddington, ed.). Edinburgh University Press, Edinburgh, Vol. 4, 1972. pp. 68–82.

Thom, R. Stabilite Structurelle et Morphogenese. Benjamin, New York, 1972. Translated as Structural Stability and Morphogenesis (see below).

Thom, R. A global dynamical scheme for vertebrate embryology. AAAS 1971, some mathematical questions in biology 4. In Lectures on Mathematics in the Life Sciences 5. American Mathematical Society, Providence R.I. 1973. pp. 3–45.

Thom, R. Phase transitions as catastrophes. Conference on Statistical Mechanics, Chicago, 1971.

Thom, R. On singularities of foliations. International Conference on Manifolds, Tokyo University, 1973.

Thom, R. Langage et catastrophes: elements pour une semantique topologique. In Dynamical Systems. Proceedings of the Symposium at Salvador, Brazil, 1971. (M. M. Peixoto, ed.). Academic Press, London and New York, 1973. pp. 619–654.

Thom, R. De l'icone au symbole: esquisse d'une theorie du symbolisme, Cahiers Internat. Symbolisme 22–23, 85–106, 1973.

Thom, R. Sur la typologic des langues naturelle: essai d'interpretation psycholinguistique. In Formal Analysis of Natural Languages, Editions Moutin, Paris, 1973.

Thom, R. Modeles mathematiques de la morphogenese. Editions 10–18, UGE, Paris, 1974.

Thom, R. La theorie des catastrophes: etat present et perspectives. Manifold 14, 16 23, 1973; also in Dynamical Systems Warwick 1974. Lecture Notes in Mathematics 468 (A. Manning, ed.). Springer, Berlin and New York, 1975. pp. 366–372.

Thom, R. La linguistique, discipline morphologique exemplaire. Critique No. 322, 235–245, 1974.

Thom R. Structural Stability and Morphogenesis (translated D. H. Fowler). Benjamin-Addison Wesley, New York, 1975. Translation of Thom, R. Stabilite Structurelle et Morphogenese with additional material.

Thom, T. Gradients in biology and mathematics, and their competition. AAAS 1974, some mathematical questions in biology VII. In Lectures on Mathematics in the Life Sciences 6, American Mathematical Society, Providence, R.I., 1975.

Thom, R. D'un modele de la science a une science des modeles. To appear.

Thom, R. Answer to Christopher Zeeman's reply. In Dynamical Systems—Warwick 1974. Lecture Notes in Mathematics 468 (A. Manning, ed.). Springer, Berlin and New York, 1975. pp. 384–389.

Thom, R. Introduction a la dynamique qualitative. Asterisque 31, 3–13, 1976.

Thom, R. Catastrophes et equations quasilineaires. In Rencontre de Cargese sur les Singularites et leurs Applications (F Pham, ed.). Institut d'Etudes Scientifiques de Cargese, Publ. Math. Dept., Univ. Nice, 1975 pp. 89–90.

Thom, R. The two-fold way of catastrophe theory. In Structural Stability, the Theory of Catastrophes and Applications in the Sciences. Lecture Notes in Mathematics 525 (P. J. Hilton, ed.). Springer, Berlin and New York, 1976, pp. 235–252.

Thom, R. Structural stability, catastrophe theory, and applied mathematics. SIAM Review 19, 189–201, 1977.

Thom, R. and Sebastiani, M. Un resultat sur la monodromie. Invent. Math. 13, 90–96, 1971.

Thom, R. and Zeeman, E. C. Catastrophe theory: its present state and future perspectives. In Dynamical Systems—Warwick 1974. Lecture Notes in Mathematics 468 (A. Manning, ed.). Springer, Berlin and London, 1975. pp. 366–389.

Thompson, J. M. T. Instabilities, bifurcations, and catastrophes. Phys. Lett. A 51, 201–203, 1975.

Thompson, J. M. T. Catastrophe theory in elasticity and cosmology. In Rencontre de Cargese sur les Singularites et leurs Applications. (F. Pham, ed.) Institute d'Etudes Scientifiques de Cargese, Publ. Math. Dept., Univ. Nice, 1975. pp. 100–104.

Thompson, J. M. T. Designing against catastrophe. 3rd International Congress on Cybernetics and Systems, Bucharest, 1975.

Thompson, J. M. T. Experiments in catastrophe. Nature, Lond. 254, 392–395, 1975.

Thompson, J. M. T. Catastrophe theory and its role in applied mechanics. In 14th International Congress on Theoretical Applications of Mathematics, Delft. 1976. North Holland, Amsterdam To appear.

Thompson, J. M. T. and Gaspar, Z. A buckling model for the set of umbilic catastrophes. Preprint, Engineering Department, University College, London, 1977.

Thompson, J. M. T. and Hunt, G. W. A General Theory of Elastic Stability. Wiley, London and New York, 1973.

Thompson, J. M. T. and Hunt, G. W. Dangers of structural optimization. Engng Optimization 1, 99, 1974.

Thompson J. M. T. and Hunt, G. W. Towards a unified bifurcation theory. J. Appl. Math. Phys. (ZAMP) 26. 581–604, 1975.

Thompson. J. M. T. and Hunt, G. W. A bifurcation theory for the instabilities of optimization and design. In Mathematical Methods in the Social Sciences (D. Berlinski, ed.). Synthese, to appear.

Thompson, J. M. T. and Shorrock, P. A. Bifurcational instability of an atomic lattice. J Mech. Phys. Solids 23, 21–37, 1975.

Thompson, J. M. T. and Shorrock, P. A. Hyperbolic umbilic catastrophe in crystal fracture. Nature, Lond. 260, 598–599, 1976.

Thompson, J. M. T., Tulk, J. D. and Walker, A. C. An experimental study of imperfection-sensitivity in the interactive buckling of stiffened plates. In Buckling of Structures, IUTAM Symposium, Cambridge Mass., 1974 (B. Budiansky, ed.). Springer, Berlin and New York, 1976. pp. 149–159.

Thompson, M. Class, caste, the curriculum cycle and the cusp catastrophe. In Rubbish Theory. Paladin, London, to appear.

Thompson, M. The geometry of confidence: an analysis of the Enga *te* and Hagen *moka*; a complex system of ceremonial pig-giving in the New Guinea highlands. Preprint, Portsmouth Polytechnic, 1973; also in Rubbish Theory. Paladin, London, to appear.

Tougeron, J.-C. Ideaux de Fonctions Differentiables. Springer, Berlin and New York, 1972.

Trinkaus, H. and Prepper, F. On the analysis of diffraction catastrophes. J. Phys. A Math. Gen. 10, 1977.

Trotman, D. J. A. and Zeeman, E. C. Classification of elementary catastrophes of codimension $\leqslant 5$. In Structural Stability, the Theory of Catastrophes, and Applications in the Sciences. Lecture Notes in Mathematics 525 (P. J. Hilton, ed.). Springer, Berlin and New York, 1976. pp. 263–327.

Ursell, F. Integrals with a large parameter: several nearly coincident saddle-points. Proc. Cambridge Philos. Soc. 72, 49–65, 1972.

Varchenko, A. N. Newton polyhedra and estimates for oscillatory integrals, Funkcional. Anal. i Prilozen. 10, 13–38, 1976.

Varchenko, A. N. Zeta-function of monodromy and Newton's diagram. Invent. Math. 37, 253–262, 1976.

Waddington, C. H. (ed.) Towards a Theoretical Biology, 4 vols. Edinburgh University Press, Edinburgh, 1968–1972.

Waddington, C. H. A catastrophe theory of evolution. Ann. N. Y. Acad. Sci. 231, 32–42, 1974.

Walgate, R. Rene Thom clears up catastrophes. New Scientist 68, 578, 1975.

Walker, W. The analysis of sudden reversals of predator–prey data. Preprint, Auckland University, 1976.

Wall, C. T. C. (ed.) Proceedings of the Liverpool Singularities Symposium. Lecture Notes in Mathematics 192. Springer, Berlin and New York, 1971.

Wall, C. T. C. (ed.) Proceedings of the Liverpool Singularities Symposium II. Lecture Notes in Mathematcis 209. Springer, Berlin and New York, 1971.

Wall, C. T. C. Introduction to the preparation theorem. In Proceedings of the Liverpool Singularities Symposoium. Lecture Notes in Mathematics 192 (C. T. C. Wall, ed.) Springer, Berlin and New York, 1971, pp. 90–96.

Wall, C. T. C. Stratified sets: a survey. In Proceedings of the Liverpool Singularities Symposium. Lecture Notes in Mathematics 192 (C. T. C. Wall, ed.). Springer, Berlin and New York, 1971. pp. 133–140.

Wall, C. T. C. Lectures on C^∞ - stability and classification. In Proceedings of the Liverpool Singularities Symposium. Lecture Notes in Mathematics 192 (C. T. C. Wall, ed.). Springer, Berlin and New York, 1971. pp. 178–206.

Wall, C. T. C. Regular stratifications. In Dynamical Systems–Wawick 1974. Lecture Notes in Mathematics 468 (A. Manning, ed.). Springer, Berlin and New York, 1975. pp. 332–344.

Wassermann, G. Stability of Unfoldings. Lecture Notes in Mathematics 393. Springer, Berlin and New York, 1974.

Wassermann, G. (r, s)-stability of unfoldings, Preprint, University of Regensburg, 1974.

Wasserman, G. Stability of caustics· In Rencontre de Cargese, sur les singularites et leurs Applications (F. Pham. ed.). Institut d'Etudes Scientifiques de Cargese, Publ. Math. Dept., Univ. Nice. 1975. pp. 128–132.

Wassermann, G. (r, s)-stable unfoldings and catastrophe theory. In Structural Stability, the Theory of Catastrophes, and Applications in the Sciences. Lecture Notes in Mathematics 525 (P. J. Hilton, ed.). Springer, Berlin and New York, 1976. pp 253-262.

Wassermann, G. Stability of unfoldings in space and time. Acta. math. 135, 57-128, 1975.

Wassermann, G. Classification of singularities with compact abelian symmetry. Regensburger Mathematische schriften 1, Department of Mathematics, University of Regensburg, 1977.

Whitney, H. The general type of singularity of a set of $2n$–1 smooth functions of n variables. Duke Math, J. 45, 220–283, 1944.

Whitney, H. The singularities of smooth n-manifolds into $(2n-1)$-space, Ann. Math. 62, 247–293, 1955.

Whitney, H. Mappings of the plane into the plane. Ann. Math. 62, 374–470, 1955.

Whitney, H. Elementary structure of real algebraic varieties. Ann. Math. 66, 545–556, 1957.

Wilson, A. G. Catastrophe theory and urban modelling: an application to modal choice. Environ. Planning A 8, 351–356, 1976.

Wilson, A. G. Nonlinear and dynamic models in geography: towards a research agenda. Working Paper 160, School of Geography, University of Leeds, 1976.

Wilson, A. G. Towards models of the evolution and genesis of urban structure. Working Paper 166, School of Geography, University of Leeds, 1976.

Wilson, A. G. Equilibrium and transport system dynamics. Working Paper 171, School of Geography, University of Leeds, 1976.

Wilson, F. W. Smoothing derivatives of functions with applications. Trans. Amer. Math. Soc. 139, 413–428, 1969.

Woodcock, A. E. R. Discussion paper: cellular differentiation and catastrophe theory. Ann. N. Y. Acad. Sci. 231, 60–76, 1974.

Woodcock, A.E.R. Embryology, differentiation, and catastrophe theory. Manifold 15, 17–33, 1974.

Woodcock, A. E. R. The development of biological form: toward an understanding? Preprint, Williams College, Mass., 1976.

Woodcock, A. E. R. and Poston, T. A Geometrical Study of the Elementary Catastrophes. Lecture Notes in Mathematics 373. Springer, Berlin and New York, 1974.

Woodcock, A. E. R. and Poston, T. A higher catastrophe machine. Proc. Cambridge Philos. Soc. 79, 343–350, 1976.

Zakalyukin, V. M. A versality theorem. Funkcional. Anal. i Prilozen. 7, 28–31, 1973.

Zakalyukin, V. M. On Lagrange and Legendre singularities. Funkcional. Anal i Prilozen. 10, 26–36, 1976.

Zeeman, E. C. Breaking of waves. In Symposium on Differential Equations and Dynamical Systems. Warwick 1968–69. Lecture Notes in Mathematics 206 (D. R. J. Chillingworth, ed.). Springer, Berlin and New York, 1971. pp. 272–281.

Zeeman, E. C. The geometry of catastrophe. Times Literary Supplement, 10th December 1971. pp. 1556–1557.

Zeeman, E. C. Differential equations for the heartbeat and nerve impulse. In Towards a Theoretical Biology (C. H. Waddington, ed.). Edinburgh University Press, Edinburgh, 1968–72. Vol. 4, pp. 8–67; also in Dynamical Systems. Proceedings of the Symposium at Salvador, Brazil, 1971 (M. M. Peixoto, ed.). Academic Press, New York and London, 1971. pp. 683–741.

Zeeman, E. C. C^{∞}- density of stable diffeomorphisms and flows. Proceedings of the Symposium on Dynamical Systems, Southampton University, 1972.

Zeeman. E. C. An essay on dynamical systems. Report to the SRC on the 1968 1971 Programme of Differential Equations at University of Warwick, 1972.

Zeeman, E. C. A catastrophe machine. In Towards a Theoretcal Biology (C. H. Waddington, ed.). Edinburgh University Press, Edinburgh, Vol. 4, 1972. pp. 276–282.

Zeeman, E. C. Catastrope theory in brain modelling. Int. J. Neurosci. 6, 39–41, 1973.

Zeeman, E. C. Applications of catastrophe theory. Manifolds, Tokyo, 1973.

Zeeman, E. C. On the unstable behaviour of stock exchanges, J. Math. Econ. 1, 39–49, 1974.

Zeeman, E. C. Catastrophe theory: a reply to Thom. Manifold 15, 4–15, 1974; also in Dynamical Systems–Warwick 1974. Lecture Notes in Mathematics 468 (A. Manning, ed.). Springer, Berlin and New York, 1975. pp. 373–383.

Zeeman, E. C. Research ancient and modern. Bull. Inst. Math.Appl. 10, 272–281, 1974.

Zeeman, E. C. Primary and secondary waves in developmental biology. AAAS 1974. Some mathematical questions in biology VIII. In Lectures on Mathematics in the Life Sciences 7. American Mathematical Society, Providence, R. I., 1974. pp. 69–161.

Zeeman, E. C. Levels of structure in catastrophe theory. Proceedings of the International Congress of Mathematics, Vancouver, 1974. pp. 533–546.

Zeeman, E. C. Differentiation and pattern formation. Appendix to Cooke, J. Some current theories of the emergence and regulation of spatial organisation in early animal development. A. Rev. Biophys. Bioengng. 4, 210–215, 1975.

Zeeman, E. C. Catastrophe theory in biology. In Dynamical Systems–Warwick 1974. Lecture Notes in Mathematics 468 (A. Manning, ed.). Springer, Berlin and New York, 1975. pp. 101–105.

Zeeman, E C. Catastrophe theory, Preprint. University of Warwick. 1975.

Zeeman, E. C. Catastrophe theory. Scient. Am. 234, 65–83, 1976.

Zeeman, E. C. A mathematical model for conflicting judgements caused by stress, applied to possible misestimations of speed caused by alcohol. Br. J. Math. Statist. Psych. 29, 19–31, 1976.

Zeeman, E. C. The umbilic bracelet and the double cusp catastrophe. In Structural Stability, the Theory of Catastrophes, and Applications in the Sciences. Lecture Notes in Mathematics 525 (P. J. Hilton ed.). Springer, Berlin and New York, 1976. pp. 328–366.

Zeeman, E. C. Prison disturbances. In Structural Stability, the Theory of Catastrophes and Applications in the Sciences. Lecture Notes in Mathematics 525 (P. J. Hilton, ed.). Springer, Berlin and New York, 1976. pp. 402–406.

Zeeman, E. C. Gastrulation and formation of somites in amphibia and birds. In Structural Stability, the Theory of Catastrophes and Applications in the Sciences. Lecture Notes in Mathematics 525 (P. J. Hilton. ed.). Springer Berlin and New York 1976. pp. 396–401.

Zeeman, E. C. Euler buckling. In Structural Stability, the Theory of Catastrophes, and Applications in the Sciences. Lecture Notes in Mathematics 525 (P. J. Hilton, ed.). Springer, Berlin and New York 1976. pp. 373–375.

Zeeman E. C. Brain modelling. In Structural Stability, the Theory of Catastrophes, and Applications in the Science. Lecture Notes in Mathematics 25. (P. J. Hilton, ed.). Springer, Berlin and New York, 1976. pp. 367–372.

Zeeman, E. C. Catastrophe theory. Proc. Roy. Instn. in press.

Zeeman, E. C. Application de la theorie des catastrophes a l'etude du comportement humain. To appear.

Zeeman, E. C. Duffing's equation in brain modelling. To appear in Bull. Inst. Math. Appl.

Zeeman, E. C. Catastrophe Theory Selected Papers (1972–1977). Addison-Wesley Reading, Mass., 1977.

Zeeman, E. C. A. catastrophe model for the stability of ships. Preprint, University of Warwick 1977, to appear in Proc. Esc. Lat.-Am. Math. 3, (1976). IMPA. Rio de Janeiro, Brazil.

Zeeman, E. C. Hall, C., Harrison, P. J., Marriage, H. and Shapland, P. A model for institutional disturbances. Br. J. Math. Statist. Psych. 29, 66–80. 1976.

Author Index

Subject Index